Essentials of Digital Construction

Lessons learned from digital transformation

Amador Caballero

Published by Emerald Publishing Limited, Floor 5,
Northspring, 21–23 Wellington Street, Leeds LS1 4DL.

ICE Publishing is an imprint of Emerald Publishing Limited

Other ICE Publishing titles:
Digital Twins in the Built Environment: Fundamentals, principles and applications
Qiuchen Lu, Xiang Xie, Ajith Parlikad, Jennifer Schooling and Michael Pitt. ISBN 978-0-7277-6580-2
BIM in Principle and in Practice, Third edition
Peter Barnes. ISBN 978-0-7277-6369-3
Asset Management, Second edition: Transforming asset dependent businesses
Edited by Chris Lloyd and Michael Corcoran. ISBN 978-0-7277-6143-9

A catalogue record for this book is available from the British Library

ISBN 978-1-83549-446-2

Cover photograph: Funtap/Shutterstock

Commissioning Editor: Michael Fenton
Content Development Editor: Cathy Sellars

Typeset by KnowledgeWorks Global Ltd.
Index created by Madelon Nanninga-Fransen

Essentials of Digital Construction

Contents

Acknowledgements

The research presented in this publication has greatly benefitted from meaningful conversations and collaborative efforts over the years with numerous colleagues and industry professionals. I am deeply grateful for the valuable experiences, lessons learned and insights they have shared, which have enhanced my understanding of the challenges and necessary actions for the implementation of digital construction.

I would like to extend my heartfelt thanks to my family, especially my wife, for their unwavering support. While I may not have been able to dedicate the time you all deserved as I worked on this book during weekends and holidays, your encouragement has been invaluable to me. I also offer my sincerest gratitude to the friends who have supported me throughout this journey, particularly those who are no longer with us to witness the fruits of their support.

About the author

Amador Caballero is a experienced professional with a deep passion for professional development in the construction industry. Holding a architectural technologist degree, a master's degree in health and safety and diplomas in 3D architectural visualisation and data analytics, he brings a multifaceted skill set to his work.

His career began on construction sites, working for main contractors, which formed the foundation of his professional experience. He has also worked in architectural practices and currently serves as enterprise architect. Over the years, he has amassed a wealth of expertise, offering him profound insights into the sector.

Amador's journey into digital construction took a significant turn in 2014 when he started leading the implementation of building information modelling (BIM), in accordance with PAS 1192-2, within a main contractor firm. Driven by a strong commitment to disruptive innovation, and with a pragmatic approach, he has focused on streamlining processes, implementing digital tools and analysing data to identify beneficial emerging trends. His extensive involvement in a multitude of BIM projects has endowed him with a comprehensive understanding of the BIM landscape in the UK.

Previously serving as the head of digital construction, Amador led the charge in implementing digital strategies and technologies, including the roll-out of BIM within the business. In his current role as an enterprise architect, his focus is on aligning technology and processes with business objectives to enhance efficiency and client satisfaction.

He has been instrumental in helping companies achieve significant milestones, such as BIM Level 2 certification from the Building Research Establishment (BRE) in 2016 and ISO 19650 1-2 certification from the British Standards Institution (BSI) in 2020. His efforts have been acknowledged through a number of UK national awards for successful BIM implementation and nominations for training strategies.

A consistent feature of his career has been his proactive involvement in seminars and discussions centred on digital construction and data analytics. He is deeply committed to supporting, educating and sharing knowledge and experiences with both supply chain partners and clients. As a fervent advocate for digital construction, Amador promotes a culture of ongoing education and learning, which he views as crucial for achieving collective goals and advancing the industry.

Preface

The purpose of sharing these perspectives is to offer a thought-provoking exploration of various subjects, drawing from my individual observations, research and experiences.

The thoughts and opinions shared within these pages are solely my own, shaped by my unique experiences, and should be interpreted as such. They are not intended to represent the official stance or endorsement of any organisation I am currently or have previously been affiliated with.

A primary aim in writing this publication is to provide individuals and companies interested in digital construction with a practical handbook that answers a number of common questions arising around digital transformation and to facilitate the effective delivery of digital transformation projects. Drawing on extensive experience in delivering digital construction projects, I aim to offer candid reflections and insights to help in avoiding the pitfalls and challenges that I, and others, have encountered.

The book serves as a practical guide for digital construction leaders and organisations, providing a clear understanding of the process of digital transformation and how to navigate it effectively. The content is grounded in my own triumphs and setbacks, offering valuable advice for successfully leading an organisation's digital evolution.

Rather than focusing exclusively on theories and standards, this book delves into the potential obstacles and significant opportunities within digital construction. I share my experiences and lessons learned while tackling such issues as resistance to change, inconsistent documentation or unreliable suppliers.

The contents of this book are a valuable resource for anyone, irrespective of their experience level in digital construction. The book is designed to help in avoiding common mistakes and navigating the different challenges that are likely to crop up.

You will probably find familiar scenarios and obstacles while reading these pages. My goal is to equip you with the tools to manage your digital construction projects with greater efficiency and confidence.

I am committed to providing an honest, unfiltered account of my experiences in digital construction. Real-world insights are more valuable than theoretical concepts alone. The aim is to deepen your understanding of the practical challenges and successes in this field, assisting you in your own successful journey in digital construction.

Empowering leaders in the digital transformation journey

This book seeks to provide valuable support to leaders embarking on the exciting and fulfilling path of leading a business's digital transformation. The journey involves acquiring knowledge about various departments within a company to standardise and improve processes, integrate new technology and enhance operations for ultimate success. Although the role can be challenging, requiring significant mental resilience to overcome resistance and opposition, it also offers a unique opportunity for growth and accomplishment.

As a leader responsible for implementing digital construction within a company, you must ensure that the organisation is digitally equipped to meet market demands. At the same time, you should drive cultural and procedural changes for sustainable and thriving transformation in the long term. It is true that your efforts might not always be acknowledged or appreciated; however, by educating businesses about the benefits of digital construction and advocating for recognition and support, you can create an environment conducive to successful implementation.

In this role, challenges and insecurities are common, and high levels of initiative, independence and adaptability are required to navigate the ever-evolving industry landscape. Overcoming both internal and external resistance can be difficult but, by sharing experiences and insights, leaders can be more prepared to handle these obstacles more effectively. Witnessing individuals embrace digital construction, challenge their preconceived notions about it and advocate for its integration is rewarding and paves the way for success in today's business environment.

By the end of this book, you will feel more confident and better equipped for managing digital construction. The book will help you to gain a deeper understanding of the necessary cultural shifts and learn how to manage requirements at various stages of a project's lifecycle. Ultimately, this publication aims to serve as a practical companion, empowering you to face challenges head-on and succeed in your particular digital transformation journey.

Progress yet to be made in building information modelling implementation

Focusing on the implementation of building information modelling (BIM) in the construction industry, this book explores BIM's significance as a cornerstone of digital construction. The BIM Mandate, published in 2011 as part of the UK Government Construction Strategy, required 'fully collaborative 3D BIM (with all project and asset information, documentation and data being electronic) as a minimum by 2016.' However, my experience suggests that the industry is still not meeting this target, owing to a lack of adequate knowledge and skills. While some early adopters, such as design consultants and large main contractors, have embraced BIM, a significant portion of the industry still needs to adopt this process for it to become standard practice.

At the time of writing, the most recent National Building Specification (NBS) *Digital Construction Report* is from 2021. Although 2 years have passed since its publication, the findings can nonetheless help us understand the current state of BIM adoption. The report, (Bain and Hamil, 2021), which surveyed 906 construction professionals, primarily from the UK but with international representation.

The survey results are concerning, revealing a limited understanding of the BIM process. For instance, 29% of respondents thought that BIM is solely about using 3D parametric models. Only 60% used a common data environment for collaboration, and merely 37% had utilised a task information delivery plan in the previous year. These percentages contradict the 71% of respondents who claimed to have adopted BIM.

Clearly, a significant portion of the industry is yet to grasp the basics of digital construction. The level of implementation varies widely; some companies are ahead of the curve, while others are lagging behind.

Rapid advances in technology are underway in the construction industry, posing a challenge for businesses to keep pace. Companies that fail to adapt to these changes risk falling behind. Therefore, rather than awaiting guidance from competitors, it is vital that companies proactively embrace these shifts. Although early adoption may present challenges, it holds the potential to provide a significant competitive edge in the long run.

The law of diffusion of innovations, conceptualised by Professor Everett Rogers (1962) and further expanded by Geoffrey Moore (1991) in his work *Crossing the Chasm*, serves as a framework for understanding the adoption of new ideas and technologies over time. This theory categorises adopters into five groups: innovators, early adopters, early majority, late majority and laggards, as illustrated in Figure 0.1. Applying this framework to BIM, we can assess its level of acceptance within the industry.

In his influential TED talk *How Great Leaders Inspire Action*, Simon Sinek (2009) discusses the law of diffusion of innovations and refers to the 15–18% penetration level as a crucial tipping point for mass market success and acceptance of an idea. This percentage serves as a useful benchmark in understanding when an innovation like BIM might start achieving broader acceptance.

Figure 0.1 Law of diffusion of innovations (adapted from Rogers, 1962)

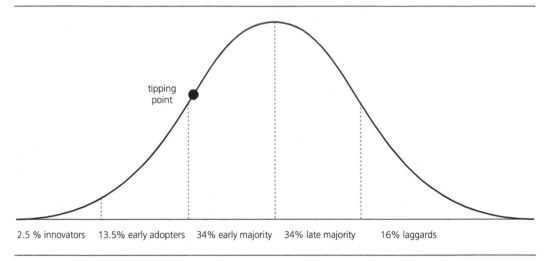

tipping point

| 2.5 % innovators | 13.5% early adopters | 34% early majority | 34% late majority | 16% laggards |

Although the NBS *Digital Construction Report* suggests that BIM may have reached beyond the initial market penetration phase, this is just the starting point. The real challenge lies in ensuring that BIM is adopted correctly and extensively, not only by major firms but also by clients and smaller companies. A high rate of BIM adoption does not necessarily translate into its effective use. It's essential for the construction industry to not only embrace BIM in terms of quantity but also to focus on the quality and resilience of its implementation. This involves continuous efforts to integrate BIM practices effectively, ensuring that they are not just widely used, but also used in a way that maximises their potential benefits in the sector.

In conclusion, the construction industry must invest in education and training to enhance the understanding and adoption of BIM and other digital technologies. By focusing on correct implementation and fostering a culture of early adoption, the industry can succeed in its digital transformation, benefiting all stakeholders. Large contractors and smaller subcontractors alike need to work collectively towards this aim, propelling the industry towards a more efficient and sustainable future.

Who this book is for

This book is designed to benefit a broad spectrum of stakeholders involved in shaping their companies' futures. While it is particularly useful for those in specialist digital roles, it is equally relevant for senior managers and directors looking to transform their businesses. Additionally, those involved in the actual delivery of projects

– including design consultants, subcontractors and clients – will find value in its insights.

The broad range of subjects that are covered should present a comprehensive resource for anyone in the built environment sector seeking to understand and apply digital construction. The content has also been designed to be accessible for those with varying levels of familiarity with digital transformation concepts.

A key strength of this resource is the grounding in real project experiences. This practical focus will enable readers to learn from actual case studies, helping them to navigate the range of issues and challenges that they may encounter.

The emphasis throughout is on actionable insights for improving work efficiency and effectiveness, thereby ensuring the content's relevance and applicability for the reader.

Practitioners

Practitioners, such as architects, engineers, surveyors or specialist contractors, typically use BIM tools for tasks such as design, analysis, coordination and actual construction.

For practitioners, BIM isn't just theoretical; it's an essential part of their daily work. The book provides insights into real-world case studies and addresses common challenges, serving as both an introduction for newcomers and a reference guide for experienced professionals. This aims to enhance individual productivity and project success. Chapters 3, 7, 8 and 9, along with Sections 4.3, 6.2, 6.6 and 6.7, are specifically designed to address the needs of practitioners.

Senior managers and directors

Senior managers and directors are influential decision makers, responsible for steering their organisations' digital transformation. While they may not directly engage with digital construction tools and processes, they oversee strategic planning, make pivotal business decisions and safeguard the company's interests.

This book offers a top-down view of how digital transformation can improve the construction process, enhance productivity and streamline operations. It emphasises the importance and benefits of initiating a cultural shift within an organisation, equipping senior managers with foundational knowledge for informed discussions and decision making. Senior managers and directors will find Chapters 1, 2 and 5, as well as Sections 4.2, 7.3 and 7.12, particularly beneficial.

Digital construction leaders

Digital construction leaders drive digital construction strategy and transformation within an organisation. They bridge the gap between new processes and solutions, integrate technology across departments, provide training and support and ensure adherence to industry standards and best practices. They are especially focused on leveraging and understanding innovations in the construction sector.

For digital construction leaders, this book serves as a comprehensive guide to digital construction, the catalyst for digital transformation in the construction industry. The book offers recommendations, practical tips and highlights emerging trends. This equips digital construction leaders to align digital construction strategies with broader organisational objectives and prepare for future technological advances. While a digital construction leader may find something valuable within all the chapters, I believe that Chapters 1, 2, 3, 6 and 8 will provide them with some useful resources to support implementation within their businesses.

Clients

In the context of this book, 'client' refers to the parties who either commission and finance the construction project or manage the information function on behalf of the appointing party.

For clients, gaining a deep understanding of the capabilities and benefits of BIM is essential for eliminating misunderstandings and garnering support for its implementation in their projects. The book provides actionable guidance on the steps to take, starting from the tender stage, to establish correct documentation and select appropriate teams. With this knowledge, clients can make well-informed decisions at different stages that align with their project goals and maximise the value of their investments. Chapters 1 and 5, as well as Sections 3.3, 6.3, 6.4, 7.10, 7.11, 7.12 and 7.13, are designed to serve as a helpful resource for clients.

Glossary

This glossary contains terms that are specifically pertinent to the content presented in this book. While this independently compiled selection offers focused insights, another, more extensive, collection of industry terminology can be found in the BRE Group's compilation of BIM terminology, available on their website (BRE Group, 2023).

Acceptance Acceptance refers to a stakeholder's formal agreement that provided information aligns with the stipulated project requirements, a responsibility carried out by the appointing party.

Accountability   The obligation of individuals and teams to take responsibility for their actions, decisions and performance and to answer for the outcomes.

Authorise   The process of providing permissions for information utilisation is a responsibility carried out by the lead appointed party.

Appointed party   The party responsible for supplying the specified information for the project, as outlined in the appointment agreement.

Appointing party   The appointing party, also known as the client or asset owner, is the recipient of information from the lead appointed party.

Appointment   The appointment pertains to the professional service agreement between parties. These are the appointing party, consultancy services and subcontractors.

Approval   The validation of information to ensure adherence to specified requirements before moving to the next stage. Approval relates to ensuring managed and shared information meets defined standards.

Artificial intelligence (AI)   Artificial intelligence (AI) utilises hardware and software to execute functions that would normally demand human cognitive skills (McKinsey & Company, 2023). The objective of AI is to streamline processes, enhance safety and improve efficiency at various stages of a building's lifecycle.

Asset information   The data collected and managed about an asset throughout its lifecycle, encompassing technical specifications, current condition and maintenance history. Such consolidated information is crucial for making informed decisions about the asset's construction, operation and maintenance.

Authorisation   The process of providing permissions for information utilisation is a responsibility carried out by the lead appointed party.

BCF   The BIM Collaboration Format (BCF) enables seamless communication between BIM tools by utilising previously exchanged Industry Foundation Classes (IFC) models. By adhering to open standards, the BCF ensures that model issues can be shared easily, thereby eliminating the need for proprietary systems.

BSRIA stages   The Building Services Research and Information Association (BSRIA) is a non-profit organisation that operates through membership. Its aim is to share knowledge and offer expert services to those involved in construction and building services. Initially released in 1994, BSRIA Guide BG 6, *A Design Framework for Building Services*, was created to provide a clear understanding of roles and responsibilities during the design stages of construction projects. Over time, the guide has been updated to keep pace with evolving practices in the UK construction sector and subsequent changes in design duty assignments.

CAFM Computer-aided facility management (CAFM) software helps facility managers manage various aspects of facility management, such as maintenance, space planning, asset tracking, operational services and related financial information.

CDP package The contractor's design portion (CDP) package comprises a set of documents that outline the contractor's design responsibilities in a project. It specifies the design deliverables expected from the contractor.

COBie The Construction Operations Building Information Exchange is a dataset designed for collecting and transferring asset information during the different stages of the asset's lifecycle. It consists of a series of tables that outline the information required for each asset within the building.

Cost of error The financial impact of mistakes made during the construction process. Poor information management, ineffective communication and inappropriate team selection are directly related to these costs.

Cyber Essentials A UK Government-backed certification scheme that helps organisations defend themselves against common cyber threats (National Cyber Security Centre, 2023). It outlines basic technical controls for cyber hygiene. The scheme comes in two levels: Cyber Essentials, based on self-assessment, and Cyber Essentials Plus, which includes hands-on technical verification. Both aim to bolster defences and reduce vulnerability to attacks.

Delivery team The delivery team includes the lead appointed party and their respective task teams or appointed parties.

IFC Industry Foundation Classes (IFC) is a standardised digital representation of the built environment, including buildings and infrastructure. Recognised as an international standard under ISO 16739-1:2018, IFC is vendor-neutral and can be used across various devices and software for different purposes. Developed by buildingSMART International (2023) to support openBIM, the IFC schema provides a systematic data model for describing the use, construction and operation of a facility. This encompasses everything from building components to work schedules and cost analysis.

Information container   A uniquely identified set of information, such as a model, drawing, document, table or schedule, that can be retrieved from within a file, system or application storage hierarchy.

Information model   A mix of structured and unstructured data, which can include geometric details, alphanumeric content and documentation.

ISO 19650   An international standard for managing information across the whole lifecycle of a built asset through the use of building information modelling (BIM). Currently consists of five parts:
   Part 1: Concepts and principles (BSI, 2019)
   Part 2: Delivery phase of the assets (BSI, 2021)
   Part 3: Operational phase of the assets (BSI, 2020a)
   Part 4: Information exchange (BSI, 2022)
   Part 5: Security-minded approach to information management (BSI, 2020b).

ISO 27001:2022   An international standard for information security (ISO, 2022), BSI (2023) for the UK, that provides a set of standardised requirements for an information security management system (ISMS) (NPSA, 2023).

Lead appointed party   The lead appointed party is directly appointed by the appointing party and is responsible for coordinating information between various task teams as the leader of the delivery team.

Machine learning   Machine learning (ML) is a form of AI that enables systems to learn from data. These systems can identify patterns, make predictions and recommendations and automatically enhance operational efficiency. Over time, they adapt in response to new data and experiences, thereby improving their effectiveness. This adaptability aids in making predictive or automated decisions that contribute to better project management.

Multifactor productivity   Multifactor productivity (MFP) assesses how well labour and capital are used in production. This includes such factors as management and scale. If gross domestic product (GDP) grows but inputs stay the same, the growth is due to MFP. This is tracked as an index and in yearly growth rates.

Project team   The project team includes everyone involved in the project, regardless of their appointment; this also includes the appointing party.

RIBA stages The Royal Institute of British Architects (RIBA) Plan of Work (2020) organises the planning, designing, building and management of construction projects into eight stages. Each stage specifies the outcomes, essential activities and required information exchanges:

Stage 0: Strategic Definition

Stage 1: Preparation and Briefing

Stage 2: Concept Design

Stage 3: Spatial Coordination

Stage 4: Technical Design

Stage 5: Manufacturing and Construction

Stage 6: Handover

Stage 7: Use.

Stages 0 to 4 are generally completed in order. Stages 4 and 5 overlap in the project schedule for many projects. Stage 5 starts when the contractor takes control of the site and ends on practical completion. Stage 6 begins with the building being handed over to the client immediately after practical completion and concludes at the end of the defects liability period. Stage 7 starts concurrently with Stage 6 and continues throughout the building's lifespan.

Scope of work A document that defines the tasks to be performed by the appointed party. It outlines the comprehensive series of tasks, responsibilities and deliverables required.

SFG20 The benchmark for specifications in building maintenance. The SFG20 application helps in creating and customising maintenance plans to align with a property's business needs. This ensures that businesses can easily stay compliant with evolving regulations.

Supply chain For the purpose of this book, the term 'supply chain' includes both design consultants and subcontractors.

Task team Those responsible for carrying out a particular task. In this book, the term 'task team' is used interchangeably with 'appointed party'. The service may be provided by either a design consultancy or a subcontractor (supply chain).

Uniclass 2015 In the UK, the construction industry has traditionally used the Common Arrangement of Work Sections (CAWS) as a classification system. However, with CAWS no longer being updated, Uniclass is increasingly being adopted for projects. The Uniclass classification is a comprehensive and unified system, consisting of various tables designed to meet different industry needs. This versatility enables facility managers and owners to classify their assets with Uniclass and helps designers and constructors organise specifications and manage projects more efficiently.

Validation Checking for compliance with permissible values, ensuring completeness and correct format; the aim is to ensure that the information and service meet the needs and expectations of users. For example, the digital team is responsible for validating the asset data.

Verification The process of confirming, through objective evidence, that specific requirements have been met and that information is accurate. This involves ensuring that information and service match the designated specifications. For example, the technical team is responsible for verifying the asset data.

Figure 0.2 is based on ISO 19650-2 (BSI, 2021) and serves as a visual aid to facilitate understanding. It shows how various parties interact, using terminology that corresponds to the ISO 19650 series.

Figure 0.2 Relationships within project team (based on BSI, 2021; Icons: Leremy/Shutterstock)

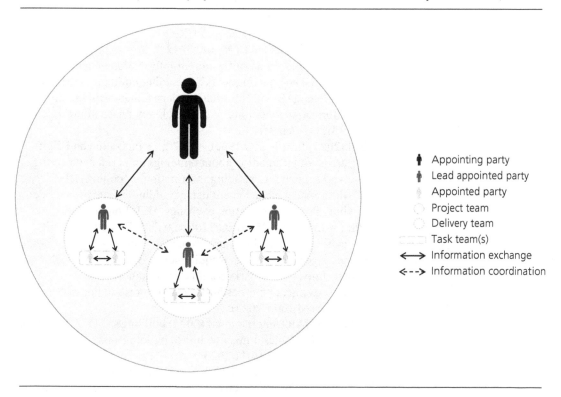

BIBLIOGRAPHY

Bain D and Hamil D (2021) *Digital Construction Report 2021*. NBS Enterprises, Newcastle upon Tyne, UK.

BRE Group (2023) BIM Terminology. https://bregroup.com/expertise/bim/bim-terminology/ (accessed 16/11/2023).

BSI (2019) BS EN ISO 19650-1:2018 Organization and digitization of information about buildings and civil engineering works, including building information modelling (BIM). Information management using building information modelling. Part 1: Concepts and principles. BSI, London, UK.

BSI (2020a) BS EN ISO 19650-3:2020: Organization and digitization of information about buildings and civil engineering works, including building information modelling (BIM). Information management using building information modelling. Part 3: Operational phase of the assets. BSI, London, UK.

BSI (2020b) BS EN ISO 19650-5:2020: Organization and digitization of information about buildings and civil engineering works, including building information modelling (BIM). Information management using building information modelling. Part 5: Security-minded approach to information management. BSI, London, UK.

BSI (2021) BS EN ISO 19650-2:2018 & Revised NA: Organization and digitization of information about buildings and civil engineering works, including building information modelling (BIM). Information management using building information modelling. Part 2: Delivery phase of the assets. BSI, London, UK.

BSI (2022) BS EN ISO 19650-4:2022: Organization and digitization of information about buildings and civil engineering works, including building information modelling (BIM). Information management using building information modelling. Part 4: Information exchange. BSI, London, UK.

BSI (2023) BS EN ISO/IEC 27001:2023. Information security, cybersecurity and privacy protection. Information security management systems. Requirements. BSI, London, UK.

BSRIA https://www.bsria.com/uk/consultancy/project-improvement/bg-6-design-framework-for-building-services/ (accessed 16/11/2023).

buildingSMART International (2023) buildingSMART Standards & Technologies. https://technical.buildingsmart.org/standards/ (accessed 16/11/2023).

ISO (2022) ISO/IEC 27001:2022: Information security, cybersecurity and privacy protection. Information security management systems. Requirements. BSI, London, UK.

McKinsey & Company (2023) What is AI? *McKinsey & Company*, 24 Apr. https://www.mckinsey.com/featured-insights/mckinsey-explainers/what-is-ai (accessed 16/11/2023).

Moore GA (1991) *Crossing the Chasm: Marketing and Selling Disruptive Products to Mainstream Customers.* HarperBusiness, New York, NY, USA.

National Cyber Security Centre (2023) About Cyber Essentials. https://www.ncsc.gov.uk/cyberessentials/overview (accessed 16/11/2023).

NPSA (National Protective Security Authority) (2023) Data Centre Security: Further Resources. https://www.npsa.gov.uk/data-centre-security-further-resources (accessed 16/11/2023).

RIBA (2020) RIBA Plan of Work. https://www.architecture.com/knowledge-and-resources/resources-landing-page/riba-plan-of-work (accessed 16/11/2023).

Rogers EM (1962) *Diffusion of Innovations*. Free Press of Glencoe, New York, NY, USA.

SFG20 What is the SFG20 Standard? https://www.sfg20.co.uk/what-is-sfg20 (accessed 16/11/2023).

Sinek S (2009) How Great Leaders Inspire Action. *TED Talks*, Sept. https://www.ted.com/talks/simon_sinek_how_great_leaders_inspire_action?language=en (accessed 22/11/2023).

Uniclass 2015 https://www.thenbs.com/our-tools/uniclass (accessed 16/11/2023).

Amador Caballero
ISBN 978-1-83549-446-2
https://doi.org/10.1680/iceedc.9446201

Chapter 1
Introduction

1.1.   Introduction

In this chapter, I take a journey through the world of digital transformation in the construction industry. As you read, you will understand the importance of digital construction, discover the driving forces behind its adoption and learn about strategic approaches to achieve the full potential of this significant shift.

Citing the latest official data from the Insolvency Service, *Construction News* reported that a record-breaking 471 construction firms went into administration in May 2023 (Vogel, 2023). An analysis of the monthly insolvency statistics in England and Wales reveals that 2022 was a difficult year for the construction industry (Webster, 2023). A total of 4163 companies faced administration, marking a significant jump from 2580 in 2021. This distressing trend is fuelled by a combination of rising inflation, supply chain problems and the persistent effects of the COVID-19 pandemic. As we moved into 2023, the situation showed no signs of improvement; by May, 1825 construction firms had already entered administration. The sector is grappling with an array of challenges, from escalating costs and labour shortages to uncertainties surrounding Brexit and increased funding costs caused by rising interest rates.

These obstacles are likely to continue to exert pressure on the industry in the future. To stay resilient and retain market relevance in this challenging environment, construction companies must embrace change and adopt digital technology in their operations to increase efficiencies and productivity.

Like all other industries, the construction sector must adapt its processes to the technological advances in our ever-changing world. This chapter highlights the critical role of digital construction in today's competitive market. Building information modelling (BIM) is the main enabler, leading to enhanced efficiency, streamlined processes and improved project delivery. This change requires not just the adoption of new technologies, but also a cultural shift, driven by senior leaders. Effective collaboration across all departments is essential for standardising processes and managing information and data accurately. It is often remarked that 'Insanity is doing the same thing over and over and expecting different results.'

Given the challenges facing the construction industry, coupled with exceedingly narrow profit margins, even minor mistakes can result in significant financial losses. As a result, businesses need to adopt digital construction methods to mitigate these issues. Digitalisation offers various benefits, including attracting skilled labour, improving productivity and efficiency, optimising supply chain management and advancing sustainability. Innovative technologies, such as 3D modelling, information management, data analytics and artificial intelligence, enable businesses to build resilience and adapt to ever-changing challenges.

This chapter explores the significance of assessing your business's current situation and objectives. This involves understanding your team's challenges and fostering a mindset towards digital transformation.

Having the right management support is fundamental in implementing a digital construction strategy and in effectively adopting technology and processes to overcome challenges that emerge throughout the implementation. Also, I highlight the importance of a cultural transition towards becoming a learning-focused organisation and holding teams accountable for successfully implementing the strategy.

1.2. What is digital transformation in the construction industry?

Rapid advances in technology are reshaping both our lives and the business landscape across all sectors. For companies to stay viable and preserve competitiveness, adopting digital transformation and finding new operational methods is not an option. Hence, it is necessary to embrace the opportunities that digital transformation and new operational methods can bring.

The significance of digital transformation cannot be overstated. Adopting it is essential for any organisation to remain competitive. Failing to embrace digital solutions and processes puts a business at risk of lagging behind competitors and losing market share. An example of this is how new entrants, like Uber, Airbnb and Netflix in other sectors, have radically changed traditional business models and client expectations.

In this section, I will explore why understanding the concept of digital transformation is essential in today's world. Many organisations, some perhaps unknowingly, are already on this transformative journey, but lack a robust strategy and roadmap to help them to navigate the necessary changes required.

Understanding the concept of digital transformation

Much like other buzzwords that frequently circulate, the term 'digital transformation' has evolved, leading to varied interpretations depending on the audience. One definition that resonates with me comes from Salesforce (2023):

> Digital transformation is the process of using digital technologies to create new – or modify existing – business processes, culture, and customer experiences to meet changing business and market requirements. This reimagining of business in the digital age is digital transformation.

What I appreciate about this explanation is its clarity: digital transformation is not just about adopting new technology. It's about a business continuously improving and altering its processes and internal culture to provide a better client experience. It's an ongoing evolution, not a one-time event.

It is important to appreciate that digital transformation requires collective buy-in from all departments and the leadership team within an organisation. It's not something that one person or department can achieve single-handedly; it must have full support from the board of directors. To truly make a lasting change, an organisation needs to shift its culture and mindset, fostering collaboration and openness to change across all departments. This is also called cross-functional alignment.

The main goal of digital transformation is to boost productivity by making processes more efficient and delivering greater value to clients through new technologies and methods. A second key element is to make processes more effective – that is, add greater value to projects. This can help a business maintain its edge in a competitive market and create new opportunities for growth.

Sadly, many companies underestimate the scope of digital transformation, viewing it as merely a tech upgrade. This is a narrow viewpoint. Digital transformation has the power to impact every facet of an organisation – from its technology and processes to its strategy, culture and overall mindset.

When discussing digital transformation, a pressing question emerges: should innovation be customer-led or technology-led? Digital transformation is multifaceted, and so too is innovation in the realm of digital construction.

Most agree that customer-driven innovation is generally effective. Companies design products based on customer needs, catering to both the client overseeing the construction and the end users. A prime example is modular construction, a method tailored to the client's demands for efficiency, cost and sustainability.

However, there's also merit in technology-led innovation. As Steve Jobs famously said, 'People don't know what they want until you show it to them' (Reinhardt, 1998). Such techniques as point cloud surveys and AI emerged without direct customer input but have since become invaluable industry tools.

The optimum strategy might be a hybrid of both. By focusing on customer needs while staying open to groundbreaking technology, companies can offer both immediate and potentially transformative solutions.

Digital construction and the three pillars for implementation

In my view, digital construction refers to the incorporation of digital transformation strategies within the construction industry. This approach utilises processes and digital tools to enhance the lifecycle of each asset, from conception and design to delivery and maintenance.

A driving force behind digital construction is the building information modelling (BIM) process. This process significantly simplifies the transition towards a more efficient and digitally focused construction environment.

Figure 1.1 shows how BIM facilitates digital construction, serving as the central component and enabler. It's important to note that the figure represents the elements associated with BIM, and that the connections between the circles are symbolic.

Successfully implementing digital construction and driving organisational transformation requires an optimal balance of process, culture and technology, as shown in Figure 1.2. This is essential for supporting the business changes required for a successful digital transformation.

Process transformation

The adoption of BIM fundamentally alters the manner in which information is produced, shared and managed across all phases of the project lifecycle. Teams from various departments must adapt to these new methodologies, which exert a far-reaching impact on everyone engaged in the project. Moreover, the incorporation of digital construction technology opens up a wide array of opportunities. For example, techniques in digital construction can substantially enhance client

Figure 1.1 BIM as enabler of digital construction

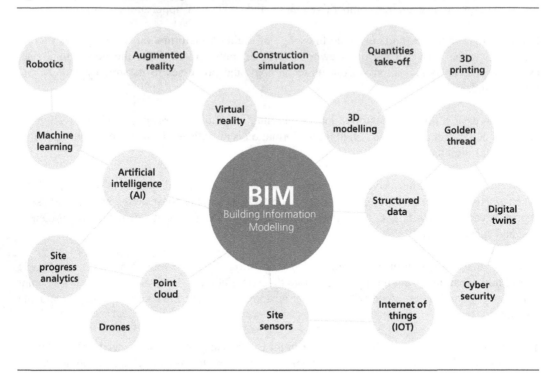

engagement through such features as virtual walk-throughs, real-time updates and other interactive experiences. Additionally, the utilisation of artificial intelligence (AI) and data analytics paves the way for more strategic decision making, streamlining not just project management tasks – such as scheduling, resource allocation and verification of the work – but also boosting the overall efficiency and productivity of a project.

Cultural transformation

In my experience, the most important and challenging aspect of digital transformation is effecting a cultural shift within the organisation. Changing tools and processes is one thing, but transforming an organisational culture and mindset is far more complex. Success relies on organisational backing and the accountability of teams for adopting new processes and utilising new tools. It is crucial that this integration takes place within the daily responsibilities of teams, with a cycle of continuous improvement fuelled by ongoing feedback. To achieve such seamless integration, a willingness to evolve is essential. This must be accompanied with appropriate training of existing staff. Moreover, replacing sceptics and recruiting new, digitally savvy talent may be necessary to achieve more of the benefits that digital transformation has to offer.

Technology adoption

In the construction sector, the introduction of new technologies can often replace existing systems, providing teams with new functionalities and simplifying tasks by phasing out obsolete tools. Advanced technologies, such as 3D design models, data analytics, AI-powered tools, machine learning and robotics, are now available to facilitate informed decision making, streamline processes and enhance efficiency and productivity. These technologies not only reduce operational

Figure 1.2 People, process and technology (based on the PPT framework; Icons: M.Style/Shutterstock)

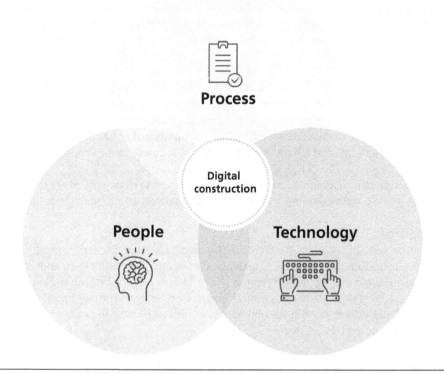

costs but also contribute to higher project quality and effectiveness. Through the implementation of digital construction techniques, businesses position themselves to outpace the competition and drive innovation.

At a time when automation is more popular than ever, the People, Process, Technology (PPT) Framework – originally introduced by Harold Leavitt in the early 1960s and later popularised by Bruce Schneier in 1999 – remains a relevant guide for organisational success (Simon, 2019). Schneier (2013) highlighted how the slow pace of human decision making could be a bottleneck, compared with swift computer operations. He advocated for automation as a solution to enhance security and efficiency. With the advent of advanced automation technologies, the PPT framework is undergoing a transformative shift. Now people need to acquire new skill sets to stay relevant, processes are becoming more streamlined than ever and technology is evolving to offer interconnected, data-driven solutions. All these elements work in tandem, making the PPT framework even more adaptable and indispensable in the current emerging innovations in the industry.

1.3.   Why we need digital construction

Historically, the construction industry has been slow to adopt new technologies and processes, relying heavily on manual methods in both design and construction. However, the adage 'modernise or die' is particularly pertinent today. In recent years, awareness has grown regarding the necessity for modernisation to remain competitive and to fulfil client and regulatory requirements.

The UK BIM Mandate (Cabinet Office, 2011), although falling short of initial expectations, acted as a significant catalyst for transformation, promoting collaboration and information sharing among stakeholders. It provided awareness of the issues that the industry is facing and the need for change. Despite these advances, the industry has yet to fully embrace technology and lean processes that drive innovation and improve performance. A considerable gap remains between the potential capabilities of new processes and technologies and the current practices of many construction organisations. To bridge this gap, the industry must proactively seize these new opportunities to boost efficiency, cut costs and elevate project quality.

Although the uptake of BIM and digital construction has been slower than is required, I remain optimistic. With a committed approach to change, the future of the construction industry seems promising, provided there is effective adoption of digital construction. This transformation could result in a healthier, more sustainable sector, reduce the number of struggling companies, preserve jobs and economic stability, and attract younger people to a career in the industry.

Digital construction has the potential to substantially enhance various critical areas currently facing challenges in the construction industry. I would like to direct attention to the following.

Skilled resources

The construction industry is currently facing a shortage of skilled workers, owing to the retirement of older generations and a lack of uptake among younger generations. To address this issue, we must modernise and change the perception of the industry by embracing digital construction methods, such as 3D models, drones, robotics, augmented reality, artificial intelligence and data analytics. Additionally, by supporting remote work, we can also make the industry more agile and appealing to attract potential new talent.

Technology is already playing a role in reducing overheads by automating tasks in the construction industry, allowing companies to deliver more work with the same number of people. However, some individuals may view technology as a threat to the profession rather than an opportunity for improvement and increased productivity. It is important, therefore, to shift the culture and mentality of the industry to embrace and support technology in order to evolve and succeed.

In the short term, the current labour shortage in the construction industry will have negative impacts. However, by attracting high-quality talent and equipping the workforce with the right skills and technology, the industry will be much better off over the longer term.

Productivity

Productivity issues are prevalent in the construction sector, frequently stemming from complex processes and inadequate focus on workforce education, development and innovation. Such shortcomings can lead to oversized teams, reducing a company's competitive edge. By automating and simplifying tasks, firms can address these productivity challenges.

In my opinion, another hindrance to productivity is the industry's persistent issue with accountability and repeated errors. A more open approach to identifying and correcting mistakes is needed, yet there is often a reluctance to confront these problems, leading to recurring issues that could have been avoided, increased disputes, compromised performance and additional rework.

Although some factors affecting productivity in construction are beyond control – such as weather conditions, client changes or unexpected site discoveries like asbestos – many can be mitigated through proper processes and the correct use of tools. Labour is often squandered on tasks that could be automated, or by resolving disputes, waiting for materials, addressing on-site issues or redoing work resulting from poor coordination and ineffective communication between different parties involved in the project.

Implementing digital construction methods can significantly improve productivity and reduce the cost of errors in projects. This is achieved primarily through improved information management. Adopting new technologies and data analytics will help address the challenges facing the construction industry, enabling better decision making, risk management and quality control.

Looking at the historical context, before the 2008 financial crisis, the construction industry was a weak spot for UK productivity, pulling down the overall productivity numbers. Since 2008, things have improved slightly in construction, with the construction industry growing faster than the rest of the economy. However, the industry has still not reached the UK average for productivity. As shown in Figure 1.3, data from the Office for National Statistics indicate that productivity in the construction industry has seen minimal improvements over the past half-century.

The construction industry is inherently unstable, influenced by global economic shifts and such events as Brexit, the COVID-19 pandemic, the Ukrainian war and inflation. These events

Figure 1.3 Output per hour worked and multi-factor productivity (MFP), UK, 1970 to 2020 (Martin, 2021; Source: Office for National Statistics – Labour productivity and multi-factor productivity) © Crown Copyright, 2021. This information is licensed under the Open Government Licence v3.0. To view this licence, visit http://www.nationalarchives.gov.uk/doc/open-government-licence/OGL

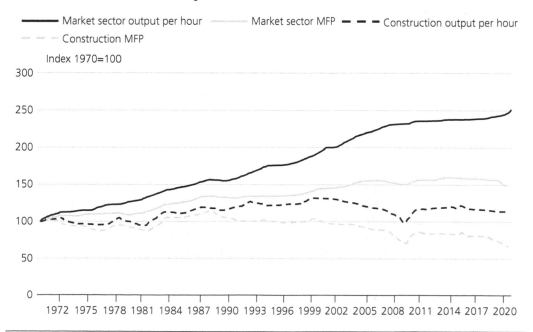

Output per hour worked and multi-factor productivity, construction industry and market sector, UK, 1970 to 2020

considerably weaken the construction supply chain, leading to project delays, elongated lead times and rising costs. Such factors can result in suppliers falling short of meeting demands, placing financial strain on construction firms. The industry operates on tight profit margins and schedules, making it difficult to absorb these disruptions, potentially leading to failure.

Given the industry's high susceptibility to external adverse events, businesses should adopt digital construction methods. These include material forecasting, real-time material tracking, planning for material shortages and off-site construction. Such measures will help organisations to be better prepared and more resilient to external pressures. Collaboration with innovative companies that employ technology to tackle these issues is necessary. By doing this, businesses can achieve improved outcomes and reduce risks.

Quality of information

Traditional construction projects have often overlooked quality information management, focusing primarily on the execution of on-site works. This neglect has resulted in rework, difficulties in retrieving information, supporting asset operation and securing accurate insurance coverage.

Digital construction, however, addresses these issues by offering better control of project information and allowing verification of the completed work against the accepted design. These capabilities ensure higher quality and build trust in the information provided to end users, as well as aiding in asset management.

This not only improves collaboration and assists construction companies in obtaining accurate insurance coverage by providing verifiable data on the quality of executed work, but it also contributes to the creation of a valuable knowledge base. This enables better decision-making for future projects. Overall, digital construction aims to bridge information gaps, helping in both project execution and compliance, thus alleviating long-standing challenges in the industry.

Sustainability

As governments intensify efforts to reduce greenhouse gas emissions, including emissions of carbon dioxide, the construction industry faces growing pressure to diminish its environmental impact and embrace sustainable practices. A key strategy is the use of technology to minimise energy consumption during construction, thereby reducing the embodied energy and overall greenhouse gas emissions of projects. The adoption of digital construction methods supports both sustainability and broader decarbonisation initiatives, contributing to a significant reduction in emissions beyond just CO_2.

Incorporating sustainable design principles and using materials with a smaller carbon footprint, such as recycled steel, is necessary. A McKinsey article (2021) underscores the substantial emissions, including those of various greenhouse gases, from processing raw materials and operational activities in construction. Efficient resource utilisation, recycling and the implementation of energy-efficient systems are key to mitigate these emissions. These strategies, coupled with digital construction techniques, play a crucial role in steering the construction industry towards a sustainable and low-carbon future, in accordance with governmental objectives to curtail greenhouse gas emissions.

Minimising waste during the construction process is essential for sustainability and cost management. Embracing modern methods of construction (MMC), which include innovative building techniques and increased use of off-site construction, can significantly reduce both material waste and a wide spectrum of greenhouse gas emissions.

While technology companies offer various solutions to cut down on greenhouse gas emissions and material waste, it's important to be both selective and practical when choosing these solutions. Investments should yield a positive return on investment (ROI) and aim beyond merely meeting targets. By making informed decisions, businesses can reduce their environmental impact without adversely affecting project costs or team productivity. However, as the 2020 Sweco report, *Carbon Cost in Infrastructure* (Barlow and Metaxas, 2020), indicates, the principle of 'reduce carbon, reduce cost' has limitations. Initially, carbon footprint reduction can lead to cost savings, but beyond a certain threshold, further reductions require additional investment. This observation adds nuance to the straightforward notion that cutting the carbon footprint always leads to financial savings.

1.4.   Strategies for maximising the benefits of digital transformation
This section highlights the challenges you may encounter and provides useful tips for implementing a digital construction strategy. It's worth noting that these tips are guidance only, and alternative approaches can also be effective. By committing to the process over the long term, construction companies can unlock the full potential of digital transformation.

Assessing current status and defining objectives
Each business's transformation journey is unique, necessitating a clear understanding of the company's current status and long-term objectives. This involves assessing the company's existing processes, services and capabilities, as well as identifying areas that require improvement or innovation, considering that the degree of transformation required may vary, depending on the services provided by the company.

While it's tempting to focus solely on the end goal, it is essential to set intermediate objectives and celebrate immediate wins; this boosts morale and maintains momentum. Regularly monitoring progress and evaluating the strategy are required steps to ensure that your transformation efforts remain aligned with the company's objectives and best practices. This also allows for flexibility in incorporating innovations and responding to lessons learned, which may necessitate adjustments to the initial approach and goals.

Before defining your digital transformation strategy, it's fundamental to consider the challenges faced by your team members, not just the views of senior management. Begin by consulting those who are directly involved in day-to-day tasks; this will give you a clearer understanding of their needs and perspectives. Collecting this feedback is crucial for shaping an informed and effective strategy, and it will also provide you with the necessary visibility within the business.

Once you have a well-rounded view, communicate the strategy from the top down. Ensure that the objectives are clearly conveyed to all departments to foster a collective understanding. Good communication is essential for securing the support of key stakeholders, such as senior leaders, directors and department heads. It's also crucial for engaging those who carry out daily tasks, as their buy-in is necessary for the successful implementation of any digital transformation strategy.

Building a culture for successful digital transformation
Success in digital transformation is closely linked to an organisation's culture, rather than just the adoption of new technologies. Leadership plays a crucial role in facilitating this cultural shift.

Organisational culture varies, necessitating tailored leadership approaches. In aiming to change organisational culture, understanding the existing culture type – be it clan, adhocracy, market or hierarchy – is relevant. Leaders must adapt their styles to guide the culture shift effectively. For instance, transitioning from a hierarchy to a clan culture may require leaders to shift from a controlling to a more collaborative and mentoring approach, ensuring alignment with the desired culture to achieve the organisation's goals. This emphasises that a one-size-fits-all approach in leadership can be counterproductive when navigating cultural change within an organisation.

The push for transformation should originate from senior leadership and cascade through every level of the organisation. Importantly, the driving force behind this effort should not be a fear of falling behind competitors or merely boosting profits, as this could lead to rash decisions. Instead, it should stem from a genuine belief in the value of change. This means that everyone, from leadership to frontline employees, needs to understand not just what is changing, but why the change is important.

Simon Sinek's (2009) concept of a golden circle provides a compelling framework for inspiring action. It's important to remember that people are motivated not just by what you do or how you do it, but by why you do it. A genuine belief in the 'why' behind the transformation can inspire others to participate in the journey and reap the benefits. This sense of purpose equips your organisation to navigate the challenges of digital transformation more effectively.

In sectors like construction, which are typically resistant to change, clear communication is necessary to overcoming scepticism. Everyone in the organisation, from senior managers to ground-level staff, must understand and support the benefits of adopting a digital strategy.

Cultural shifts within an organisation are not achieved overnight; they require time, dedication and a willingness to change attitudes and behaviours. Senior teams may find it challenging to embrace new ideas, particularly if those ideas push them out of their comfort zones. Just as Rome wasn't built in a day, implementing these changes also takes time and patience. The pace of transition largely depends on the level of support and commitment from directors and senior management.

Continuous learning and upskilling are vital for creating the right culture that enables successful digital transformation. Even while teams are occupied with daily tasks, time must be set aside for growth and adaptation. Learning from mistakes is another key aspect. Errors encountered during the transformation should not be viewed as failures but as opportunities for improvement. The organisation's ability to learn from these lessons and avoid future mistakes will be a strong indicator of the success of various initiatives and the business as a whole.

Management support and leadership

Active support from directors is crucial to overcoming cultural resistance within an organisation. Management backing is even more important than having the right strategy. A well-crafted strategy won't guarantee success unless it has the appropriate level of backing. The board must be actively engaged and understand both the value and the positive impact that digital transformation will have on the business.

Digital transformation involves change, which not everyone will accept in the same manner, despite the benefits it may bring to their roles. Without the active support and buy-in from directors, any initiative has a significant chance of failure.

Directors and senior management must spearhead the transformation process and cultural shift, setting a precedent for the rest of the organisation. It's a requisite to clearly communicate the vision behind the change, identifying the issues it aims to address, and outlining the potential benefits. The leadership team should encourage skill development and mastery of new capabilities, while setting clear expectations and holding individuals accountable for adhering to new processes. This approach is indispensable for the successful implementation and acceptance of new organisational processes.

In this way, managers will foster a positive and supportive environment, where team members feel motivated to embrace the change and contribute to the team's success. Furthermore, managers must cultivate a culture of learning and continuous improvement to ensure that the benefits of digital transformation are fully realised.

A lack of leadership commitment will reflect in the behaviour of other team members, hampering the transformation process. As a result, implementation might not be taken seriously enough by some team members, leading to poor execution and repeated mistakes. Some might even resist incorporating the transformation in their team or project requirements.

Strategic technology implementation
Implementing technology to achieve the full benefits of digital transformation necessitates a strategic and considered approach. Choose technology that aligns closely with your business needs, strategies and long-term objectives.

Every decision should be made carefully, taking into account the various technological options available and consulting the appropriate teams. A pragmatic evaluation is crucial, as is considering the needs and expectations of the end user. To avoid redundancy and ensure efficiency, your business should also establish a technology architecture that aligns closely with its requirements.

In industries like construction, where digital transformation is becoming increasingly relevant, caution is advised. While some technologies and concepts hold great promise, they might not yet be mature enough or the organisation may not be ready for such a change for full-scale implementation. Premature adoption could negatively affect your team's perception of and engagement with transformation initiatives. Therefore, it's advisable to thoroughly evaluate the maturity and suitability of a technology or concept before rolling it out across departments.

Finally, adopting a commercial mindset that views digital transformation initiatives as long-term investments, rather than short-term expenses, is crucial. This approach can secure the necessary resources and organisational support. It also allows businesses to fully understand the potential value of their efforts, considering both immediate and future returns.

Building a strong team of internal advocates
People are often afraid of the unknown, and this can lead to resistance to digital construction initiatives. However, my experience shows that negative perceptions usually improve once successful digital projects have been completed. Moreover, those involved in such projects frequently become advocates who can persuade hesitant team members.

For a higher chance of success, concentrate on assembling a strong internal advocacy team. Rather than trying to win over everyone – a draining and potentially unproductive effort – concentrate on assembling a team of individuals with the right mindset. Those who are initially resistant will eventually have to adapt if they want to fit into the evolving company culture.

A robust team of internal advocates is essential for the success of any digital transformation in the construction industry. These advocates are not only enthusiastic and committed to the initiative, but they also go above and beyond to promote it. These individuals should not only understand the benefits of digital change but also be able to help others to see them; they should be capable of driving a cultural shift within the organisation. Empower this team with the necessary resources, training and support to champion your initiatives. Their role is fundamental for effective communication between top management and the workforce, helping to ensure that the vision for digital transformation is broadly accepted and implemented.

When it comes to change management or the roll-out of new initiatives, identifying and influencing key stakeholders is essential. Each category of stakeholder within your organisation should be won over to ensure effective support for the change.

- *Gatekeepers.* These individuals control access to resources or decision making processes. Earn their support by presenting compelling evidence and a persuasive argument of the benefits of digital transformation.
- *Influencers.* These are individuals capable of altering the opinions of others. Involve them early in the process, listen to their advice and ask for their backing.
- *Blockers.* These individuals typically resist change. Engage in dialogue to understand and create strategies to address their concerns. Surprisingly, once these concerns are addressed, blockers often become some of the most passionate supporters.
- *Bystanders.* These are the undecided individuals. Illustrate how digital transformation can offer both personal and organisational growth, and they may well become valuable team members.

If you identify these key stakeholders and engage them in a strategic manner, you will be able to create a favourable environment that will facilitate a successful digital transformation.

Overcoming challenges in digital transformation

Driving the digital transformation journey is no easy task. It demands significant commitment from senior leaders and a resilient mindset from those at the helm. I find this task challenging, yet ultimately rewarding.

Leading a digital transformation can be isolating, making the unwavering support of senior managers not just helpful, but crucial for maintaining morale and focus. Navigating change requires a robust mental constitution since constant pressures and frequent resistance are common. It is a challenging task that demands specific qualities in a leader. To guide a team effectively through this complex process, leaders need a mix of important traits. They must be committed and determined to complete the project, and they need the perseverance to get past any hurdles that come their way. It's also crucial for them to have an optimistic outlook that can lift the team's spirits. A strong sense of self-belief is necessary, along with the ability to solve problems as they arise. Finally, they should be tolerant of failure, seeing it as a chance to learn rather than as a dead-end. Having these qualities can help leaders create a strong team culture that is ready to tackle the challenges of digital transformation.

When you introduce new initiatives that are foreign to the business, reactions can be mixed. While some are willing to embrace change, most remain sceptical and resistant, raising questions about the tangible benefits of such alterations. The challenge is constant: proving the value of each proposed change becomes a necessity.

Another significant barrier is erected by the individuals who openly criticise new changes, focusing solely on the potential drawbacks while ignoring or undervaluing the benefits. Maintaining composure is crucial in these situations. While it's important to consider both positive and negative feedback for improvement, being selective about which criticisms to address first and understanding the underlying reasons can prevent project derailment and keep things on course.

In any transformation process, some individuals will remain unconvinced, regardless of the strength of your arguments. My advice is to keep your emotions in check and not let detractors drain your energy. You cannot win every battle; choose the ones that warrant your main effort and concentrate on those to continue executing your strategy. Avoid getting side-tracked by pessimistic viewpoints and stay committed to your path. Change is inevitable due to industry demands, and some people will inevitably be left behind.

Even if people are successful, society often focuses solely on their mistakes. Therefore, even if your implementation brings benefits to your projects, people will tend to highlight the downsides. So today I would like to remind you that mistakes are part of the process. It is a true adage that 'A person who never made a mistake never tried anything new.' Or perhaps one could remark: 'The only man who never makes a mistake is the man who never does anything.'

Table 1.1 offers various considerations to assist you in both defining and implementing your digital construction strategy, with the aim of addressing potential challenges and maximising the benefits of digital transformation.

Table 1.1 Checklist: suggested strategies (continued on next page)

Assessing current status and defining objectives
- ☑ Conduct a comprehensive assessment of the company's existing processes, services and capabilities.
- ☑ Identify areas requiring improvement or innovation.
- ☑ Set long-term objectives for the digital transformation journey.
- ☑ Develop intermediate objectives and celebrate immediate wins.
- ☑ Regularly monitor progress and evaluate the strategy.
- ☑ Conduct internal consultations with team members at all levels to gather feedback.

Building a culture for successful digital transformation
- ☑ Secure support from senior leadership for the transformation.
- ☑ Communicate the 'why' behind the transformation to all team members.
- ☑ Foster clear communication to overcome resistance to change.
- ☑ Develop a dedicated team to facilitate the cultural shift.
- ☑ Create an environment that encourages continuous learning and upskilling.

Management support and leadership
- ☑ Gain active support and buy-in from directors and senior management.
- ☑ Assign management with spearheading the transformation process and cultural shift.
- ☑ Develop an accountability framework for teams to comply with new processes.
- ☑ Address any instances of poor execution or resistance to change.

Table 1.1 Continued

Implementing strategic technology
☑ Choose technology that aligns with business needs and long-term objectives.
☑ Conduct a pragmatic evaluation of technology options, consulting appropriate teams.
☑ Establish a technology architecture that aligns closely with business requirements.
☑ Assess the maturity and suitability of technologies before full-scale implementation.

Building a strong team of internal advocates
☑ Identify individuals within the organisation who are open to digital transformation.
☑ Empower these individuals with the necessary resources and training.
☑ Utilise internal advocates for effective communication between top management and the workforce.

Overcoming challenges in digital transformation
☑ Maintain focus and morale with the support of senior managers.
☑ Address both positive and negative feedback constructively.
☑ Prioritise challenges and criticisms that are most pertinent to the project's success.
☑ Stay committed to the transformation strategy despite resistance and scepticism.

BIBLIOGRAPHY

Barlow L and Metaxas S (2020) *Urban Insight: Carbon Cost in Infrastructure*. Sweco, Leeds, UK. https://www.swecogroup.com/wp-content/uploads/sites/2/2022/12/urban-insight-report_carbon-cost-in-infrastructure.pdf (accessed 16/11/2023).

Cabinet Office (2011) *Government Construction Strategy*. Cabinet Office, London, UK. https://assets.publishing.service.gov.uk/media/5a78ce8eed915d07d35b2933/Government-Construction-Strategy_0.pdf (accessed 23/11/2023).

Martin J (2021) *Productivity in the Construction Industry, UK: 2021*. Office for National Statistics, Newport, UK. https://www.ons.gov.uk/economy/economicoutputandproductivity/productivitymeasures/articles/productivityintheconstructionindustryuk2021/2021-10-19 (accessed 17/11/2023).

McKinsey & Company (2021) Call for action: seizing the decarbonization opportunity in construction. *McKinsey & Company*, 14 July. https://www.mckinsey.com/industries/engineering-construction-and-building-materials/our-insights/call-for-action-seizing-the-decarbonization-opportunity-in-construction (accessed 16/11/2023).

Reinhardt A (1998) Steve Jobs: 'There's sanity returning'. *BusinessWeek*, 14 May. https://web.archive.org/web/19991110003323/http://www.businessweek.com/1998/21/b3579165.htm (accessed 16/11/2023).

Salesforce (2023) What Is Digital Transformation? https://www.salesforce.com/eu/products/platform/what-is-digital-transformation/ (accessed 16/11/2023).

Schneier B (2013) People, process, and technology. *Schneier on Security*, 30 Jan. https://www.schneier.com/blog/archives/2013/01/people_process.html (accessed 17/11/2023).

Simon B (2019) Everything you need to know about the People, Process, Technology Framework. *smartsheet*, 14. June. https://www.smartsheet.com/content/people-process-technology (accessed 17/11/2023).

Sinek S (2009) *Start With Why: How Great Leaders Inspire Everyone to Take Action*. Penguin, London, UK.

Vogel B (2023) Latest Insolvency Service stats underscore grim business climate. *Construction News*, 19 Jul. https://www.constructionnews.co.uk/financial/latest-insolvency-service-stats-underscore-grim-business-climate-19-07-2023/ (accessed 17/11/2023).

Webster D (2023) *Commentary – Monthly Insolvency Statistics June 2023*. The Insolvency Service, London, UK. https://www.gov.uk/government/statistics/monthly-insolvency-statistics-june-2023/commentary-monthly-insolvency-statistics-june-2023 (accessed 16/11/2023).

Amador Caballero
ISBN 978-1-83549-446-2
https://doi.org/10.1680/iceedc.9446202

Chapter 2
Business transformation

2.1. Introduction

Given the rapid changes in the construction industry, shaped by new regulations, shifting client demands and intensifying competition, staying ahead necessitates ongoing adaptation and improvement. This chapter provides guidance to help you navigate and implement these essential shifts effectively.

The first step in this journey is to gain an understanding of the digital construction process and its impact on your business based on the services that you provide. A clear strategy is crucial: know where you are and where you want to be. This helps lay the groundwork for successful transformation. It's essential to avoid the fear of missing out (FOMO) and to make informed choices based on your organisation's needs and available resources.

In the construction industry, it can be tempting for companies to focus solely on achieving turnover targets by accepting projects that might not match their specific skill sets and resources. This preoccupation with immediate revenue generation often overshadows the necessity of streamlining internal processes and investing in the professional development of teams. Unfortunately, this short-term approach can have detrimental long-term effects, such as project failures and financial losses. Starting your strategic planning early is crucial, as waiting for a more favourable economic climate or observing competitors could leave your organisation behind in a rapidly evolving market.

Achieving success requires a harmonious blend of well-defined and standardised processes, complemented by skilled teams and accountability measures. Ignoring these key elements can lead to a loss of control and direction. Although digital technologies play a crucial role, they aren't a silver bullet; it's equally important to focus on transforming the people and processes involved.

Mistakes are inevitable, but fostering a no-blame culture with open communication transforms these setbacks into learning opportunities and helps to prevent future errors. A culture that prioritises ongoing learning not only enhances team morale but also increases productivity, thereby facilitating the easier implementation of changes. The importance of maintaining open lines of communication and a learning-first approach is invaluable.

Clear communication is essential throughout any transformation process. A thoughtfully designed communication strategy keeps stakeholders in the loop and maintains their engagement. Helping teams understand the rationale for change and its consequences is crucial for sustaining both productivity and morale. The committed involvement of the board of directors and senior management is vital for a successful transformation. Their strong backing not only ensures team accountability but also contributes to meaningful, lasting change rather than superficial adjustments.

As roles within the organisation evolve in the process of transformation, it becomes essential to upskill employees to meet new responsibilities. By fostering a culture of learning, you not only prepare your team for these evolving roles but also demonstrate a commitment to their long-term career growth. This approach enhances employee engagement and motivation, contributing to the overall success of the transformation.

To summarise, business transformation is an intricate and continuous undertaking, requiring well-thought-out planning, efficient execution and a steadfast dedication to ongoing learning. Achieving the objectives of such a transformation relies heavily on accountability, backed by support and commitment from directors and senior leaders. The following chapters will provide you with the necessary tools and insights to navigate your organisation through this multifaceted journey.

2.2. Define a strategy: where you are and what you want

The evolution of BIM adoption and its impact

In recent years, we have observed a transformation in the manner in which designers and sub-contractors approach their work. This transformation has been gradual but ongoing, characterised by a move towards the adoption of 3D design. This shift is like the one that transpired in the 1980s when designers and subcontractors started to transition from traditional drawing boards to computer-aided design (CAD). Moreover, in recent years, the industry has realised the benefits of placing more emphasis on and assigning appropriate importance to the management of information throughout a project's lifecycle.

The 2011 BIM Mandate has been a key driver for this shift (Cabinet Office, 2011). Although our current understanding and approach to BIM, with a greater focus on information management and data, has evolved from the emphasis on 3D design that was present in 2011, it created a demand for professionals with the necessary skills to drive the BIM process.

As with any new technology or process, there has been a learning curve for those involved in adopting BIM. Despite the challenges faced along the way and the adoption not being as success-ful or widespread as expected under the BIM Mandate, we have made significant progress in our understanding of how to use BIM effectively on projects.

A positive aspect of the BIM Mandate is the emergence of new technology companies aiming to support the construction industry through BIM use. These technologies have made it easier for non-technical users to unlock these benefits and increase efficiency and productivity for all project stakeholders. As a result, BIM adoption as a standard approach is becoming increasingly common within the construction industry.

Back in 2014, when I began working on BIM projects based on the PAS 1192-2:2013 standard (BSI, 2013), the situation was quite different from what we see today. Only a few designers and subcontractors had in-house capability to meet the requirements, with challenges and a steep learn-ing curve while delivering projects. These companies typically had one or two specialists, trained in 3D modelling, and BIM was not yet the norm or standard way of working within these busi-nesses. The lack of BIM implementation was also evident during the tendering process, as we would receive information in a 2D format that did not follow the appropriate naming protocol. Today, it is much more common for companies to have trained teams capable of delivering BIM projects or working towards that goal as a standard approach, with both technical and process compliance. During the tender process, it is now rare to receive information solely in the traditional 2D CAD format. Instead, most of the information we receive is presented in a 3D environment and follows the correct naming protocol.

While BIM adoption has become more widespread in recent years, some consultants and sub-contractors, particularly smaller ones without in-house designers, have been slower to embrace this approach. This is unfortunate, as it is crucial for all trades involved in the design and construction process to adopt BIM to enhance coordination and effectively manage project information. However, I believe that most of these companies, with the appropriate education and support, will adapt their processes, to remain competitive and meet the expectations of clients.

Although the adoption of BIM has grown in recent years, there is still a broad discrepancy between companies in the implementation, and understanding, of BIM. Some may claim to be BIM-compliant, with a few team members capable of creating 3D design models, but this does not necessarily indicate true capability. Therefore, clients must carefully evaluate the BIM capabilities of potential consultants and subcontractors, ensuring they work with professionals genuinely skilled in delivering project requirements.

Despite the progress, there is still much to learn about implementing BIM and digital construction efficiently and effectively in the industry. As digital construction implementation becomes more widespread, we can anticipate continued progress and innovation in this area, with professionals refining their skills and discovering new ways to leverage digital construction to improve design and construction processes. Industry-wide collaboration will be critical for driving the successful adoption of BIM and digital construction. As more companies collaborate and share their knowledge and experience, we can expect ongoing improvement in project management and delivery, leading to better outcomes for all stakeholders.

Successfully adopting BIM and digital construction

To ensure the successful adoption of BIM and digital construction, you must first thoroughly understand the BIM process. The implementation and responsibilities of BIM differ, depending on the nature of various businesses, making it fundamental to grasp the specific responsibilities and contractually obligated deliverables for each project. Bear in mind that the level of transformation required may vary based on the services your business provides. A solid understanding of the process will better equip you to integrate BIM and digital construction in your operations successfully.

Once you've grasped the process, evaluate your internal procedures to identify areas for standardisation, simplification and automation. Carefully review and improve your existing processes and explore market solutions that match your strategic objectives. Opt for solutions that truly benefit your business rather than adopting tools merely because they are popular. This approach will improve your efficiency and effectiveness in meeting both client and internal team requirements.

After gaining a clear understanding of the process and evaluating your internal operations, allocate sufficient time for the transformation and ensure support from all relevant departments. Assess your business's current position and your desired outcomes to formulate a coherent strategy for implementing BIM and digital construction. It's also crucial to secure the appropriate resources and support to avoid becoming bogged down by minor issues and maintain focus on the strategy.

Starting your transformation early is essential, since delaying it for a better economy or observing what your competitors do might put your organisation at a disadvantage in a swiftly changing market. Avoiding the need to adapt puts you at risk of missing valuable business opportunities. Accept that embracing this change will lead to greater success.

When considering the engagement of third parties for BIM deliverables due to limited internal capabilities, weigh the pros and cons. While third parties may offer specialised expertise, they can also be costlier and may result in reduced control over the process. Evaluate these trade-offs carefully before making a decision. Finally, focus on developing in-house skills and expertise for greater process control. While there may be instances when reliance on third parties is necessary, a comprehensive understanding of the BIM process is crucial for all businesses in the construction sector.

Developing a digital construction strategy

When crafting a comprehensive digital construction strategy aligned with your business model and priorities, it's crucial to consider various aspects. These include understanding the rationale, identifying challenges and opportunities and implementing effective digital transformation strategies. These are some key elements to consider when developing your high-level digital construction strategy.

Rationale

The initial step towards crafting a comprehensive digital construction strategy entails understanding why your business perceives digital transformation as a necessity. This understanding is crucial, as it helps identify the drivers and benefits that digital transformation can offer to your business. For instance, digital transformation is key for enhancing the services you provide to clients reduce the cost of error and improving overall business performance through the utilisation of digital tools and contemporary industry processes. It also aids in refining project delivery and streamlining internal operations, enabling you to stay ahead of the digital transformation curve, remain competitive and seize new market opportunities. Furthermore, staying ahead of the digital curve is instrumental in attracting and retaining top talent within the business. Additionally, digital construction assists in fulfilling client needs and meeting regulatory mandates, such as the Information Management Mandate (IPA, 2021), the Building Safety Act 2022 (HMG, 2022) and the golden thread of information (which is described in Chapter 7).

Challenges

When implementing digital construction, it is important to recognise the potential obstacles your company might face. These could include clients who are unsure of their BIM responsibilities, needs or desires, making it challenging to fulfil their requirements and meet their expectations. Insufficient or unclear documentation at the tender stage can hinder effective planning and implementation of a lean BIM process. Working with task teams who might not have the required skills or resources to fulfil their BIM duties could result in delays or issues on the project.

Moreover, internal resistance to change is not uncommon; some team members may view the BIM process and other new technologies as burdens, rather than valuable tools for improvement. This resistance to adopting new technology can impede the seamless implementation of digital construction methodologies.

Lastly, the lack of a clear strategy and vision for the technology roadmap and training can further complicate matters. Without a well-defined plan and senior leadership buy-in, it becomes difficult to align the team's efforts towards a cohesive goal. A shortfall in support from senior leaders can result in a lack of accountability across teams and compromise successful digital transformation.

Opportunities

The implementation of digital construction can provide numerous benefits to your company and teams. For instance, digital construction can enhance preconstruction performance through early stakeholder engagement, improved information management and 3D visualisation, in turn facilitating timely decision making. It can also enhance project delivery quality through improved collaborative processes and design coordination, serving as a trusted source of advice for clients and consultants, to meet their BIM expectations and support business development. Digital construction fosters innovation and new ideas from preconstruction and operations teams, increasing efficiency and productivity, and providing an operationally efficient building that showcases your services and promotes them to other buildings in your clients' estates. Moreover, it helps streamline preconstruction and construction processes, reducing slippage, defects and costs associated with subpar delivery, while also improving risk management, resulting in less rework and waste.

Implementation plan

A change of such significance, which affects the entire business, cannot be accomplished solely by one department. It necessitates support from the entire organisation. To this end, I recommend the formation of a dedicated implementation team. This team should comprise individuals who hold a suitable level of seniority and are committed to the change. Members should be drawn from diverse areas of the organisation. Their responsibilities will extend beyond merely drafting an implementation plan that meets business objectives. They will also be tasked with formulating the overarching strategy and facilitating its implementation across various departments.

Although this list is not exhaustive, here are some key areas to focus on.

- One important factor in your strategy is understanding your clients' requirements and ensuring that you can meet them. To support the development of your strategy, it's essential to be aware of your clients' preferences and expectations. This may involve gathering feedback, conducting market research and analysing client data to identify areas for improvement. Use this information to decide on and prioritise the necessary changes and actions within your business.

- Leadership is also critical to the success of your digital transformation. Your senior management team should provide clear direction and support to guide the transformation of your business. For example, leaders should set goals, hold teams accountable for their delivery and success, establish a vision for their teams and departments and support you in creating a culture that encourages the adoption of new technologies and processes.

- Engaging all departments and team members is essential to the success of your digital transformation. All employees should be competent, empowered and engaged in the adoption and implementation of digital construction. This may involve providing training, encouraging collaboration between departments, providing constructive feedback on the processes and tools implemented, using digital tools as the main means of working and fostering a culture of continuous learning.

- A process approach is also important for ensuring that your digital transformation efforts are effective. Make sure that your processes are understood, managed, compliant and regularly improved. Consider adopting agile approaches for digital transformation which allow for more flexible and adaptive workflows. This may involve identifying and eliminating bottlenecks, streamlining workflows, providing feedback on areas for improvement, complying with regulations, automating tasks with new technologies and using data to identify opportunities for improvement and capture lessons learned on previous projects, from both positive and negative outcomes.

- Furthermore, it is crucial to maintain a focus on improvement and innovation. Look for opportunities to collaborate with other organisations and drive significant improvements in a rapidly changing technological environment. This may involve piloting new technologies, experimenting with new processes and seeking out partnerships that can help you to improve your performance and stay ahead of the curve.

Figure 2.1 suggests some areas within your business that should be carefully considered as part of your overall strategic planning process.

Figure 2.1 Areas of impact: illuminating the reach of your strategy

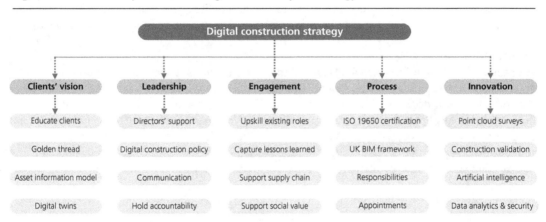

By addressing these key areas, you will be better equipped to develop a comprehensive digital construction strategy that effectively tackles the challenges, leverages the potential benefits and ensures the successful implementation of digital transformation within your organisation.

2.3. Don't get FOMO; choose your strategy wisely

The Cambridge Dictionary (2023) defines 'fear of missing out' (FOMO) as a 'a worried feeling that you may miss exciting events that other people are going to, especially caused by things you see on social media'. This phenomenon has become increasingly prevalent in the digital age, characterised by the constant sharing and broadcasting of experiences and activities.

In recent years, the construction industry has witnessed a surge of technology companies aiming to introduce modern efficiencies and boost productivity. As part of this digital transformation, it is crucial to carefully select tools that align with your business processes in order to drive efficiency.

New technologies often disrupt existing business practices. Before adopting them, it's crucial to standardise your current processes. The usual approach would be to standardise, simplify and then automate with the appropriate tool; however, in certain situations, you might consider standardising your processes in tandem with adopting the new tools, rather than waiting. Regardless of the approach, having a standardised process is necessary. However, choosing the right tools is not always a simple task. The challenge arises from balancing the expectations of various stakeholders in the business, each of whom may have differing opinions on the best technology for their specific needs.

FOMO is a genuine issue in the construction industry and can lead to hasty decision making. The pressure to stay current with the latest trends and technologies can be overwhelming, particularly when senior leaders observe competitors investing in cutting-edge tools. The temptation to make swift decisions based on social media, competitor feedback or superficial sales pitches and misleading marketing may result in poor long-term growth, excessive spending and, ultimately, under-utilised technology and loss of credibility. The goal should be to choose tools that are compatible with your business processes and offer genuine value, rather than merely following the latest trend.

It's important to remember that a technology that succeeds for one business might not yield the same results for another. This discrepancy can be attributed to the impact of internal processes, organisational structures and each company's unique requirements on the technology's appropriateness.

It's not uncommon for some senior members without appropriate experience to attempt to influence the selection of tools. However, when considering the implementation of new technology in your business, it is fundamental to approach the decision making process with a clear and focused mindset. Do not let external pressures sway your choices, and ensure you allocate sufficient time to conduct comprehensive research, understand your team's needs and analyse the options to determine what will work best for your organisation. Keep in mind that you will be held accountable for the outcome of your decision, so it's important to make a choice that aligns with your diligence and convictions.

By carefully evaluating new technology options and taking into account the unique needs of your organisation, you can avoid falling victim to FOMO. Adopting this thoughtful approach allows you to make informed decisions that genuinely benefit your company by enhancing efficiency and productivity and fostering long-term growth. The aim is to identify and implement the most appropriate technology that meets your organisation's specific requirements and objectives, not to embrace the latest trends solely because they are new.

Selecting the right tool

To guarantee the successful selection and implementation of new technology in any business, adopting a user-centric approach is fundamental, as is taking a systematic and comprehensive method. Begin by clearly defining not only the business problem you aim to resolve but also the users' needs. By including users' perspectives and requirements in your business case for each scenario, you create a more holistic approach to solution selection. This dual focus not only aids in identifying tools and solutions that are precisely tailored to your operational needs but also ensures that they match users' expectations, thereby avoiding the pitfalls of overly complex or short-lived technology.

Companies lacking a clear strategy often find themselves grappling with an excess of platforms, leading to duplication and inefficiencies. Having a roadmap can guide you in simplifying the array of tools you're using. Before diving into new technology acquisitions, seek opportunities to replace or consolidate existing systems, to prevent overlap and waste.

It is important to understand that, on occasions, people think that acquiring a specific tool will fix their issues; that is not the case. Before making a decision, it is vital to standardise and simplify processes throughout the organisation. People within the same company might follow different methods to achieve similar outcomes, resulting in potential conflicts and inefficiencies. Establishing a clear and uniform process that everyone agrees on and follows will ensure a strong foundation for the successful integration of new technology.

All systems in a business are interconnected in some way. When selecting a new tool, it's essential to understand its potential impact on existing systems. Rarely will a new addition operate in isolation; it often influences other processes and tools. Choose judiciously, keeping the broader ecosystem in mind. Keep in mind that no single tool can address all problems or meet all requirements. Integrating new technologies with existing systems typically necessitates an application programming interface (API), so it is advisable to avoid closed systems that do not permit such interaction. Hence, work with companies that are open to sharing their APIs and collaborating with others, as this will facilitate integration and improve overall efficiency. Conduct thorough research and comparison of different options, engage with departments that will benefit from the new technology and gather feedback from teams to ensure everyone is on board with the decision.

It is fundamental to consult with your information technology (IT) department during this process to guarantee that the new technology is compatible with existing systems, complies with the security protocols and requirements and can be seamlessly integrated in your business processes. An overarching enterprise architecture perspective is necessary. This ensures that different technologies integrate seamlessly, eliminating duplications in processes and tools, thus avoiding confusion and resistance among end users.

When investigating a new technology, carefully consider its benefits to the end user and its potential to enhance productivity. The successful adoption of a technology depends not only on its utility and benefits but also on its ease of use.

After selecting the tool, ensure that there is a well-thought-out implementation plan. Underestimating the importance of ongoing support and training from the solution provider can lead to setbacks in technology adoption. A good implementation should offer continuous support and training, tailored to different roles within the organisation, rather than a one-size-fits-all approach, to ensure that the tool is being utilised to its full potential.

In addition to user-friendliness, obtaining the necessary support from senior leaders is crucial for the successful deployment and adoption of new technology. People are often resistant to change, and even the most efficient and user-friendly technology can face opposition from certain employees. Therefore, senior leaders should be actively involved in the process, ensuring that the tool is being used effectively and addressing any challenges that might arise during implementation.

Collaboration between different departments and stakeholders is essential, as is involving them in the decision making process. Teams are more likely to embrace new technology if they have had a say in the selection process. This approach will minimise potential challenges and ensure a smoother adoption of the chosen technology, fostering a sense of ownership and commitment among employees.

Successfully deploying the right tool within a business is a complex undertaking, and making the wrong choice can have significant consequences, not only for the efficiency and productivity of the business but also for the credibility of those responsible for selecting and implementing the technology. Consequently, it is essential that every effort is made to ensure that the right technology is chosen and deployed effectively.

As already mentioned, FOMO is a common issue that can negatively impact the technology selection process. To mitigate the effects of FOMO, be mindful of its influence and approach the technology selection process with caution. This means taking the time to thoroughly evaluate the business needs and processes, and making a well-informed decision about the appropriate tool for the job. Refrain from selecting tools based on competitors' actions or social media trends, and focus on your company's unique needs and processes.

Additionally, when considering the implementation of new technology, be mindful of the future roadmap of your business. Understand how the new tool will work with upcoming changes, potential duplications or gaps in your processes. Ensure that the implementation process is well-planned and executed, with minimal disruption to normal business operations.

Table 2.1 is designed to guide you through the key considerations in making informed technology selections. Aligning your choices with the unique needs and objectives of your organisation will drive operational efficiency and help you achieve your strategic goals.

Table 2.1 Checklist: informed technology selection (continued on next page)

General planning
- ☑ Clearly define the problem you aim to resolve or improve.
- ☑ Develop a well-structured business case for each scenario.
- ☑ Evaluate your organisation's unique needs and requirements.

Process standardisation
- ☑ Review your process and evaluate the necessity of eliminating non-beneficial steps.
- ☑ Simplify complex or redundant steps in your current processes.
- ☑ Confirm that standardised processes are uniformly followed across departments.

Stakeholder alignment
- ☑ Identify key stakeholders in your organisation to be involved in the selection process.
- ☑ Gather their opinions on the technologies under consideration.
- ☑ Secure necessary support from senior leaders for technology adoption.
- ☑ Establish clear governance to guide decision-making and technology selection.

Tool selection
- ☑ Research potential tools that could resolve or improve the defined problem.
- ☑ Compare the pros and cons of different technology options.
- ☑ Check the potential impact of new tools on existing systems.
- ☑ Evaluate tools for their compatibility with standardised processes.
- ☑ Prioritise tools that offer seamless integration capabilities, especially through APIs.

IT consultation
- ☑ Consult your IT department for compatibility with existing systems.
- ☑ Ensure that the new technology complies with security protocols and requirements.
- ☑ Adopt an overarching enterprise architecture perspective to ensure holistic integration.

User experience
- ☑ Assess the tool's ease of use and benefits to the end user.
- ☑ Allocate sufficient time for team training.
- ☑ Plan for ongoing support to address future challenges and updates.

Table 2.1 Continued

Avoiding FOMO (fear of missing out)
- ☑ Conduct a risk assessment to identify any FOMO-driven decisions.
- ☑ Avoid making choices based solely on competitors' actions or social media trends.
- ☑ Review your company's long-term goals and how the tool aligns with them.

Implementation and adoption
- ☑ Create an implementation plan that addresses the specific needs of different user groups.
- ☑ Involve teams in the implementation process for better acceptance.
- ☑ Address any challenges that may arise during implementation.

Postimplementation review
- ☑ Assess the effectiveness and efficiency of the new tool.
- ☑ Measure key performance indicators (KPIs) after implementation.
- ☑ Review any setbacks in technology adoption and address them promptly.

In conclusion, to successfully implement new technology in any business, regardless of size, requires to standardise and simplify processes, ensure user-friendliness and obtain the necessary support from senior leaders. Following these principles will help overcome challenges, lead to smoother adoption and maximise the benefits of integrating new tools and solutions in your organisation.

2.4. Implementation policy

Policy implementation and benefits

To promote the adoption of digital construction and reduce resistance to its implementation, I recommend establishing an internal policy supported by your company's board of directors. This policy should apply to all projects regardless of whether the client has specific requirements for digital construction. The aim is to equip your teams with the skills they need to implement the correct procedures and reap the benefits, in much the same way that the military focuses on continuous training to maintain peak readiness. This approach will also ensure that your teams are prepared for clients who strongly wish to include digital construction in their projects and will be capable of meeting their expectations successfully.

This initiative requires a dynamic process of learning, continuous monitoring and evaluation to assess benefits and identify any shortcomings. By continuously implementing the digital construction policy, you can address any deficiencies that are discovered along the way and make adjustments as needed. This approach will help you to naturalise the process within your teams and lead to new discoveries that you might not have considered in the past. These discoveries might come in the form of errors that cause headaches, but they also provide valuable lessons that are not recorded in books to support you in future projects.

The policy should reflect a strong commitment to fostering a digital construction culture within the organisation, promoting continuous improvement, collaboration, innovation and efficiency. A digital construction culture promotes a collaborative environment where everyone is encouraged to embrace new technologies, embrace change and think outside the box. The policy also supports the creation of a culture of innovation, which drives the development of new processes that help to streamline and optimise construction projects. This leads to increased productivity, reduced costs and faster project delivery.

Implementing a comprehensive internal policy on digital construction can offer significant benefits to your company. This policy won't just sustain momentum in adopting digital tools and processes; it will also showcase your company's dedication to integrating digital construction in your project objectives. Both your employees and your clients will clearly see this commitment, reinforcing the idea that your company embraces digital construction and follows good industry practices for project delivery.

By making digital construction a standard practice in your business, you will ensure that all team members are familiar with the process and responsibilities and have a consistent level of knowledge. This is essential for the effective and efficient deployment of digital construction in your projects. With everyone on the same page, your teams can work together seamlessly and deliver the best results possible.

Moreover, implementing digital construction consistently on every project, not just when requested, is crucial to maintaining the skills and confidence of your teams. Digital construction is a constantly evolving field, and large gaps in implementation can result in your teams falling behind in terms of skills and knowledge. By keeping a consistent approach, your teams will be able to keep up with the latest best practices and be better equipped to deliver outstanding results.

It so happened, years ago, that as the implementation of digital construction became more widespread in the company where I was working as head of digital construction, it became quite evident that a consistent approach was necessary to ensure that every team was aligned with the business's vision and mission. A policy was needed to make sure that the teams were aware of their roles and responsibilities within the digital construction process and that they understood the benefits that digital construction could bring to their projects. Also, it was important to address and overcome the initial resistance from some individuals within the company towards adopting this process in their projects, although all the necessary elements were in place to reap the benefits of digital construction.

Board of directors' support

To be effective, a policy must be supported by the board of directors, who would then be able to promote its implementation throughout the business. This support is crucial in helping to eliminate any resistance to digital construction that might exist within the teams, and it will help to create a culture of continuous improvement and collaboration within the company. Without the active support of key stakeholders, the transformation of the business will not succeed, leading to increased frustration among those responsible for leading the implementation of the policy.

Reflecting on my experiences in my previous role, we tackled the issue of inconsistent approaches by developing a new BIM policy. This policy established that digital construction was to be implemented in every project from the beginning, regardless of whether it was part of the client's contractual requirements or not. This was an important step in ensuring that the teams were familiar with the digital construction process and that there was a consistent level of knowledge within the company. The goal was to provide as much exposure to digital construction as possible, so that the team could build confidence and gain knowledge. This would also allow for capturing lessons learned, to improve the overall implementation in projects and the business.

Special cases and exceptions

The digital construction policy needs to be comprehensive; I recommend including a specific exception regarding the implementation of building information modelling (BIM) in projects. This exception comes into play when a client's architectural and engineering information is already at RIBA Stage 3 (Spatial Coordination) without adhering to the proper BIM process. In such cases, the digital construction policy – which mandates that the BIM process be implemented in every project, regardless of the client's requirements – would not apply. This exception arises not from a lack of interest in implementing BIM, but because the BIM process should start at the outset of a project. Implementing BIM retroactively on reaching Stage 3 could lead to substantial costs and delays, owing to the necessity of revising completed design work, a challenge the project might not be poised to tackle.

I recommend that this exception be reviewed and assessed on a project-by-project basis, as the benefits of BIM, in the long run, may outweigh the costs of redoing the initial Stage 3 design. The goal is to provide teams with the flexibility they need to make informed decisions that align with the business's vision and project possibilities.

The success of the policy

Looking back on my experiences as head of digital construction, when we implemented the BIM policy, a pivotal moment in our company's journey towards digital integration, I can see that it was a transformative period that marked the beginning of a new era in our approach to construction projects. The success of the BIM policy was evident in the significant rise in the number of digital construction projects, providing more opportunities for teams to gain exposure to digital construction processes and tools. This allowed them to become familiar with their roles and responsibilities, leading to a more seamless delivery of project requirements. Moreover, the teams that had experienced the benefits of digital construction in their projects were more likely to support its implementation in future projects, as they saw the advantages that it provided. This helped to create a virtuous cycle of success, as each team's positive experiences with digital construction encouraged other teams to embrace it as well.

Personally, the digital construction policy was an important step in supporting the company's digital transformation goals. It provided the teams with a consistent approach to digital construction and helped to create a culture of continuous improvement and collaboration within the organisation. However, to achieve its full potential and drive the desired change across the business, encouragement from the board of directors is crucial. Their active endorsement is key to overcome any resistance from individuals or teams that may be reluctant to change in their work processes. It is important to note that, without this support, even the most well-crafted policy might struggle to achieve its intended impact.

I recommend that the business conducts internal audits, as they are invaluable for keeping teams vigilant and ensuring that there is no deviation from established policy. It is crucial for the business to support these audits and establish appropriate consequences for non-compliance. These audits act as a catalyst for maintaining process quality and for initiating internal reviews.

If you have third-party certification, annual audits will be conducted to ensure that business performance is in accordance with the relevant ISO 19650 standards (BSI, 2019, 2020a, 2020b, 2021, 2022). Teams generally respond well to the mention of audits, as no one wants to be responsible for non-compliance or risk losing the business certification.

Core principles of the policy

The following set of principles can serve as a comprehensive guide to assist you in defining the principles of your policy. By shaping expectations and providing clear direction, these principles can lay a solid foundation for your policy framework. However, it is required to customise and adapt these guidelines to align perfectly with your specific business expectations, objectives and unique needs.

- *Compliance and standards.* Prioritise adherence to industry standards, such as the BS EN ISO 19650 series, and follow the guidelines for the UK BIM Framework. This ensures consistency and compatibility across projects, both present and future.

- *Collaborative environment.* Utilise technology to cultivate a collaborative environment, enhancing communication throughout the project lifecycle. Specify that the use of a common data environment (CDE), approval workflows and naming conventions is mandatory for every project.

- *Simplification.* Commit to simplifying processes by eliminating unnecessary complexity and streamlining operations. This includes identifying and removing activities that don't add value, reducing the duplication of tools and consolidating the location of internal project information to save time and prevent information loss.

- *Technology roadmap.* Continuously explore, understand and implement suitable technologies to optimise performance at different stages of the project. Align these explorations with technology roadmaps to prevent duplication and ensure relevant benefits for the business.

- *Continuous improvement and learning.* Promote a culture of ongoing improvement and learning within the organisation by acknowledging successful practices and identifying opportunities for enhancement by embracing digital construction technologies. Encourage employees to adopt a mindset of embracing mistakes without fear, enabling collective learning and ensuring that errors are avoided in future projects.

- *Accountability and responsibility.* Demonstrate commitment to fulfilling roles, being accountable and ensuring the successful execution of digital construction projects. Adhere to the clearly defined roles, responsibilities and decision-making authority assigned to different management levels.

- *Competence and training.* Prioritise equipping team members and task teams with the skills needed to perform well in a digital environment. To achieve this, offer comprehensive support, along with targeted training. Share lessons learned from previous projects, to ensure the confident and effective use of digital tools and processes.

- *Data integrity and integration.* Understand the importance of reliable data for making informed decisions and driving successful outcomes. To ensure the highest data quality, teams should be held fully accountable for collecting, validating and verifying data throughout every project. This holistic approach ensures data integrity across the entire project lifecycle, aligning with organisational information requirements (OIR).

- *Data-driven decision making.* Actively analyse project data to uncover valuable insights and trends, enabling better-informed decisions and driving continual improvement in operations.

- *Product tracking and traceability.* Place an importance on the ability to track and trace all procured and installed products. This approach strengthens control over the supply chain, guaranteeing quality and compliance in all aspects of projects.

- *Embracing automation.* Actively support the adoption of automated platforms to bring efficiencies and streamline processes throughout the procurement, design, and construction stages.

2.5.   The value of lessons learned

Capturing lessons learned, both good and bad, is essential for refining future projects and ensuring that successes are repeated while avoiding past mistakes. Regularly reviewing both positive and negative results provides valuable knowledge. Mistakes can provide learning opportunities to improve future projects, while successes can become examples of best practice.

When things go wrong during the implementation of digital construction in a project, it is important to take a step back and assess the situation. The first thing to do is to acknowledge and take responsibility for your actions. It's important to avoid lying to yourself and seeking excuses, and instead recognise that what you did was not good enough on this occasion. Many factors could lead to errors during project delivery. Perhaps better communication with the design consultants and subcontractors was required to understand the project requirements; perhaps the documentation was not clear enough or did not cover certain aspects in sufficient detail to prevent the error from occurring; maybe the internal team members were not held accountable enough to deliver the BIM requirements or that additional training was needed.

The next step is to evaluate the situation and determine what needs to be improved to ensure that similar issues don't occur again in future projects. This involves discussion with all internal and external stakeholders to understand the required areas of improvement. Once you have a clear understanding of the issue, you can take steps to address it and prevent it from happening again.

It's also important to hold your team accountable for their actions and make sure that everyone understands what went wrong and what needs to be done to prevent similar issues in the future. Here is where the importance of having a culture of continuous improvement and accountability and the support of senior leaders is critical; otherwise, it is likely that the team will repeat the same mistakes.

It's worth noting that errors and mistakes are an inevitable part of digital construction. Even the best laid plans can go wrong and there will be times when you are faced with unexpected challenges. When these situations arise, it's important to stay calm and focused, and to communicate the issue to your team as soon as possible. By working together, you can minimise the impact of a mistake and quickly implement solutions to mitigate any negative effects.

Remember that the implementation of digital construction on each project is a learning opportunity. By documenting the lessons learned from each mistake and sharing these with your colleagues, and the business as a whole, you can help to prevent similar issues from arising in the future. This documentation should be easily accessible to all stakeholders and updated regularly, so that everyone can benefit from the lessons learned.

The impact of repeated mistakes cannot be underestimated; not only can they lead to financial losses, but they can also have negative effects on the morale of the delivery team and the reputation of your business and your initiatives. Therefore, it is imperative to establish a strategy for dealing with mistakes and avoiding their repetition.

One of the most important aspects is the acknowledgement of mistakes and conducting a thorough analysis of what went wrong. It is a common tendency for people to try to hide, or make excuses for, errors but this will only lead to their repetition in the future. Admitting to mistakes and analysing the root cause requires courage, honesty and accountability, which are essential for any organisation to thrive.

However, not everyone might be willing to admit their mistakes, as there might be a culture of blame in the workplace. This can lead to divisiveness among team members and hinder the process of analysis, making it more difficult to find the root cause and the actions needed to improve. It is important to create a safe and secure environment in which individuals feel comfortable discussing their mistakes and collaborating with others to identify solutions.

As leadership expert and author John C Maxwell (2001) wisely said, 'A man must be big enough to admit his mistakes, smart enough to profit from them, and strong enough to correct them.'

Communicating lessons learned

Documenting errors and identifying areas for improvement are crucial, but equally important is communicating the lessons learned to the wider organisation and maintaining well-documented records. If these records can be easily accessed, they can be used to prevent similar mistakes in the future, ensuring the organisation continuously improves and progresses towards its goals.

Conducting thorough critical analyses and collaborating with stakeholders to pinpoint the root causes of mistakes will foster creative thinking and innovative solutions, contributing to the long-term success of the business. In the scenario I previously described, during my tenure as head of digital construction, my objective was to develop a method that would turn previous challenges into learning opportunities. This was to be achieved by methodically recording all issues encountered in the BIM process throughout each project. The goal was to document these in a standardised format, thereby creating a repository of knowledge that could be used to refine practices and avoid repeating past mistakes. This approach was aimed at enhancing the efficiency and accuracy of BIM processes under my leadership.

Although the production of a case study at the end of each project is valuable for communication and knowledge sharing within teams, these are often unhelpful if you need to identify and analyse common trends in project issues. To overcome this limitation, I have now adopted a standardised practice for collecting information about problems encountered and the actions required to resolve them. Presenting these results on an easily accessible dashboard, accessible to all members of the organisation, ensures that users can readily observe trends and pinpoint common issues across various projects. This approach not only enables the identification of the occurrence and frequency of specific issues over time, but also facilitates organisational growth by providing valuable insights. Furthermore, it fosters a culture of continuous improvement, encouraging team members to actively engage with the data and contribute to the evolution of our practices.

The power of the dashboard, and similar tools, lies not just in the documentation and analysis of mistakes, but also in the collaboration this encourages between departments and stakeholders. A centralised platform allows team members to view and understand critical mistakes and their impact on projects from the tender stage, and to work together to find innovative solutions that prevent the recurrence of issues, thus improving the overall delivery of projects.

The effectiveness of the dashboard hinges on the commitment, involvement and accountability of all team members in maintaining up-to-date information and consulting it during early project stages. This facilitates awareness about lessons learned from previous projects that may apply to new ones. It's crucial to educate the team about the dashboard's purpose and benefits and ensure that everyone is aware of their role in keeping it current and relevant.

Continuous improvement is a core aspect of this initiative, requiring ongoing monitoring and updates. As new projects are completed and new mistakes are identified, these findings must be included in the dashboard to keep the information up to date and relevant.

To enhance dashboard comprehension and help users identify relevant issues, mistakes can be categorised using various factors, such as survey design models, information management and commercial considerations. Problems can also be sorted by department, improving the visibility of issues pertinent to each department and their categories. Additionally, sorting mistakes by the year in which they occurred can help track trends in error reduction over time. Issues can also be classified by their level of criticality, based on the potential impact that repeating the same mistake could have on the project and the business.

Teams should not hesitate to identify issues in their reports, and a blame culture should be avoided. Instead, it's essential to recognise the problem, collaborate as a team and provide the necessary assistance to tackle any challenges the team may encounter, to eliminate repeated issues in projects. A shift in approach when discussing problems, from a blame-focused mindset to one that offers support and encourages collaboration, will foster more effective communication and problem-solving within the organisation.

In conclusion, learning from mistakes made during digital construction projects is an important part of continuous improvement. It's crucial to acknowledge mistakes, analyse the root cause and involve all stakeholders in finding solutions. A culture of accountability and honesty is vital, along with a safe and secure environment where individuals feel comfortable discussing their mistakes. A tool such as a Power BI dashboard can be used to document and analyse past mistakes, encouraging collaboration and innovation to prevent similar issues from occurring in the future. The success of the dashboard relies on the buy-in and participation of all team members and requires ongoing monitoring and updates to remain current and relevant. By continuously learning from mistakes, organisations can make progress towards their goals and improve their overall delivery of projects.

Cost of error

The cost of error and the subsequent analysis of this cost are closely tied to the value of lessons learned and the insightful use of data. Evaluating the costs of errors and understanding the impact these errors have had on your business is a critical process. It helps identify areas of improvement and supports robust risk management in your projects.

To ensure that you mitigate risk effectively and avoid repeated mistakes, it's crucial to maintain a clear understanding of the losses incurred and the reasons behind them. Capturing the appropriate lessons learned from each project and incorporating them in future project planning is a vital part of this process.

It is important to establish a standardised approach to measuring and recording these costs and their underlying causes. With this consistent methodology in place, you can track the costs of errors across all projects in a meaningful way. Accurate and regular recording of this information facilitates the understanding of where to target improvements, enabling you to enhance operational efficiency and productivity.

Remember, analysing and learning from errors offers many benefits to the business. Not only can this analysis lead to improvements in workflows, reducing the likelihood of future mistakes, but it can also save costs and improve financial outcomes. Highlighting gaps in skills or knowledge within your teams is another advantage. Moreover, persistent errors can negatively impact a company's reputation and team morale, so addressing these errors promptly is of utmost importance.

To capture the full impact of these errors, it's recommended to strive for a granular understanding of their cost. This approach entails looking beyond a high-level view and delving deeper into the detailed aspects of each error. Doing this will provide you with a more comprehensive understanding of their impact and aid in the formulation of strategies to prevent them.

In the pursuit of measuring the cost of error, consider the following categories as a starting point. However, bear in mind that these are just examples, and you should identify and include the relevant situations that apply specifically to your business.

Design-related costs

- *Incorrect information.* Additional costs arise from misinformation provided by or shared with consultants or subcontractors.
- *Poor coordination.* Lack of coordination between task teams can lead to rework and increased labour and material costs.
- *Design errors and omissions.* Costs may be incurred to correct design mistakes or to incorporate missing elements, which could potentially lead to penalties or legal expenses.
- *Design development and scope gaps.* Additional costs arise when work that is required was not initially included in the project scope.
- *Regulatory non-compliance.* Failing to meet building codes can lead to fines and extra costs.

Planning and cost allocation errors

- *Tender.* If the bid doesn't accurately account for all related costs, or if the estimates are inaccurate, this oversight could lead to unexpected additional expenses.
- *Appointment.* Leaving out costs in subcontractor agreements may result in unexpected expenses.

Rework and defect rectification costs

- *Rework.* Mistakes and failure to adhere to the agreed design can result in increased labour and material costs because of the need for dismantling and rebuilding.
- *Defect rectification.* Postconstruction inspections can reveal non-compliance with regulations, resulting in the cost of replacement and reinspection.

Costs related to delays and disruptions

- *Abortive works and visits.* Wasted trips because of poor coordination lead to extra costs. For example, a team may arrive at the site to find their work area is not ready, requiring a return visit.
- *Lack of materials and machinery.* Unavailability of essential machinery can halt construction, leading to extended rental and labour costs.
- *Project delays.* Delays in the project can lead to increased costs in various areas, such as site setup, daily operations, management and administration. Additionally, you may incur penalties for not completing the project on time.
- *Poor supply chain performance.* Delays in material delivery or subpar performance from suppliers can result in higher labour costs, the need to source from alternative suppliers and additional costs caused by project delays. Depending on the industry and specific project, this could be one of the most significant sources of error in costs.

2.6.　Communication strategy

There is often initial resistance when it comes to embracing digital construction and BIM, owing to various misconceptions and fears surrounding new processes and technology. These misconceptions can lead to reluctance from industry professionals, who may feel threatened by the need to change their established work methods or who may be sceptical about the benefits of digital construction.

It's fundamental to address these concerns and misconceptions in order to facilitate a smoother transition to digital construction practices. While it is important to acknowledge that digital construction and BIM are not a one-size-fits-all solution to every challenge in the construction industry, their implementation can undoubtedly lead to significant improvements. These advances can positively impact the industry as a whole, as well as the individuals working within it, by making their jobs more efficient and streamlined.

To achieve successful implementation of digital construction, having an effective internal communication strategy is necessary. The aim of internal communication is twofold – first, to gain robust support from top-level management and second, to generate a desire to learn within the different teams in the business.

Presenting a compelling vision of how digital construction will drive business growth, as well as highlighting the challenges and opportunities that lie ahead, can help align business leaders with your digital construction strategy. This backing is essential for overcoming the inevitable obstacles that arise during digital construction implementation within the business.

Establishing an effective communication strategy ensures that everyone in the organisation, from entry-level employees to top management, is on board with the digital construction initiative. Engaging team members by highlighting the advantages and impacts of digital construction on their day-to-day work can be achieved through various means, such as town hall meetings, workshops and targeted training sessions. These forums provide opportunities for employees to ask questions, address concerns and share ideas about the digital transformation process.

Using peer influence and success stories

I suggest starting by concentrating your communication efforts on individuals who are open to embracing digital transformation. These individuals will become your champions, promoting digital implementation within the business and inspiring others to follow suit.

Early adopters within the business are necessary, as they serve as ambassadors who can inspire others to embrace digital implementation. Sharing examples of successful adoption by colleagues within your organisation who have embraced the change and reaped the benefits is more relatable and credible to teams than a message that comes from the digital construction leader. Providing concrete examples of improved project outcomes, personal growth and professional development can make the digital construction strategy easier to sell. By showcasing these success stories, you demonstrate that digital construction is not just a theoretical concept but a practical solution that brings tangible benefits to the organisation. This approach can help generate interest, support and enthusiasm for the transition among team members and encourage more people to participate in the process. Moreover, it reinforces the notion that digital construction is not just a trend, but a critical component of the organisation's success in the long term.

By promoting the progress and benefits of digital construction both internally and externally, you can gain increased support from your teams, drive improved results and ultimately achieve your desired objectives. This approach will also help create an internal demand for and desire to learn about digital construction, fostering a culture of innovation and continuous improvement throughout the organisation.

Moreover, this internal demand will contribute to creating a learning culture that embraces change and supports ongoing professional development. Establishing mentorship programmes and learning resources, such as workshops and online courses, can further reinforce employees' commitment to digital construction.

Additionally, I have noticed that when people who were first hesitant about digital construction start embracing it in their projects, they usually become more comfortable and confident. As they become familiar with the technology and processes, they realise it's not as scary as they first thought, and they can see the real benefits for themselves. Communicating this shift in how others view digital construction can encourage more people to adopt it and create a positive attitude towards it.

As the number of successful digital construction projects grows, so do the opportunities for teams to exchange their experiences, learn from one another and refine the overall implementation process. Therefore, it is beneficial to steadily increase the number of BIM projects within the organisation, regardless of client requirements. This approach helps evangelise your teams, bolstering support for the communication strategy and ensuring teams are prepared and proficient in delivering projects when it becomes a contractual necessity.

Furthermore, this approach creates a ripple effect, leading to more advocates who will actively promote the implementation of digital construction on their projects. As a result, business transformation becomes a smoother process for all stakeholders involved. The support of the board of directors is invaluable in driving this transformation, as they can use their influence to champion digital construction adoption and inspire the rest of the organisation to follow suit. A well-designed communication strategy that highlights the benefits of digital construction, shares success stories and addresses concerns will go a long way in securing the support of the board and enabling a more digitally savvy business environment.

In conclusion, overcoming initial resistance to digital construction and BIM requires a multifaceted approach. By securing the support of top-level management, developing an effective communication strategy and fostering a culture of innovation and learning, your organisation can successfully navigate the transition to digital construction. As more individuals experience the benefits first-hand, they will become advocates for this transformation, helping to create a more competitive, efficient and successful organisation in the long run.

To support your communication strategy, consider the following suggestions.

- *Promotional videos.* Share real-life experiences and positive results from internal employees, task teams and clients who have been involved in digital construction projects. This helps to build excitement and encourage others to embrace the process and the technology.
- *Impact on roles.* Clearly communicate the benefits of digital construction for individual roles and address any misconceptions or false beliefs about these methods. By targeting the specific needs and concerns of each role, you can more effectively build support and encourage adoption.

▨ *Training sessions.* Provide bespoke training sessions tailored to each role. These sessions should be delivered by employees from the relevant department, as they can provide practical examples and insights into the benefits of digital construction. Make sure that the business prioritises attendance at these sessions and provide opportunities for follow-up and additional training as needed.

▨ *Knowledge base.* Develop an internal knowledge base that provides access to training content, news, guidance, templates and workflows related to digital construction. This can be an invaluable resource for employees who have questions about the processes and technology, or need support during a project.

▨ *Updates.* Keep employees informed and up to date by regularly sharing information through newsletters, blogs, podcasts and other communication channels. Consistent, timely updates help to build trust and encourage continued support for digital construction initiatives.

▨ *Case studies.* Share lessons learned from past digital construction projects, both the successes and the failures. This helps to build transparency and credibility and provides valuable insights for future projects.

▨ *Social media.* Utilise social media platforms to share information and build a community of supporters for digital construction. However, be mindful of the content you share and be strategic in your approach, as oversharing or sharing the wrong information can harm your reputation.

▨ *Feedback.* Encourage feedback from all stakeholders, including employees, clients, consultants and subcontractors. Listen carefully to both positive and negative feedback, and use it to continuously improve the digital construction process. Feedback is a valuable tool for making informed decisions and improving outcomes.

▨ *Visibility enhancement.* Several of the aforementioned initiatives will increase the prominence of your team and the activities being executed or in progress within the organisation. Nonetheless, maintaining visibility with your teams is crucial. Conducting site visits and being present on site are essential to foster a more personal connection and demonstrate that your department is approachable and ready to provide support when needed. Building close relationships with individuals, making yourself approachable and actively considering opinions enables a comprehensive understanding of the business. Moreover, it is important to diversify the teams you engage with to avoid relying solely on the same perspectives.

Table 2.2 serves as a practical guide, containing various ideas and considerations for formulating a communication strategy that garners full support from your organisation. The aim is to engage all departments by tailoring messages to their unique needs and interests. A one-size-fits-all approach will not resonate with different departments.

Remember that each digital construction lead must adapt these recommendations to their specific business context and requirements. By thoughtfully considering and customising these suggestions, you can effectively communicate the benefits of digital construction, gain robust support within your organisation and, ultimately, achieve improved outcomes.

2.7.   Upskill existing roles

As a passionate advocate for digital construction, I make it my primary goal as a business leader to facilitate change and promote education within team structures. I view my role as an internal consultant, striving to transform business operations and adapt to the digital age. A digital construction lead should focus on educating and guiding businesses through digital transformation, ultimately aiming to become redundant as the culture evolves and teams assume new responsibilities. However, if you

Table 2.2 Checklist: communication strategy

Gaining management support
- ☑ Secure board of directors' support.
- ☑ Present a compelling vision that outlines the benefits of digital construction.
- ☑ Highlight challenges and opportunities to align business leaders.

Communication strategy
- ☑ Develop an internal communication plan.
- ☑ Identify platforms for communication (e.g. town hall meetings, newsletters).
- ☑ Create targeted training sessions for different roles within the organisation.

Using peer influence and success stories
- ☑ Identify early adopters who are open to embracing digital construction.
- ☑ Showcase examples of successful adoption from within the organisation.
- ☑ Share concrete examples of improved project outcomes, personal growth and professional development.

Creating an internal demand for learning
- ☑ Foster a culture of innovation and continuous improvement.
- ☑ Establish mentorship programmes.
- ☑ Provide learning resources (e.g. workshops, online courses).

Monitoring employee comfort levels
- ☑ Observe how team members adapt to using digital construction in projects.
- ☑ Share shifts in perceptions to encourage more adoption.
- ☑ Steadily increase the number of BIM projects within the organisation.

Utilising various communication channels
- ☑ Create promotional videos showcasing real-life experiences.
- ☑ Clearly communicate role-specific benefits to dispel misconceptions.
- ☑ Develop an internal knowledge base with training content, news and guidance.
- ☑ Regularly share updates through newsletters, blogs or podcasts.
- ☑ Publish case studies to share lessons learned.
- ☑ Use social media strategically to share information and build support.
- ☑ Gather feedback from all stakeholders for continuous improvement.
- ☑ Enhance visibility through site visits and by engaging with various teams.

Adaptation and customisation
- ☑ Tailor the recommendations to suit your unique business context.
- ☑ Constantly update your communication strategy based on received feedback.

are a digital construction leader, don't panic. As innovations emerge and technology advances, the role will continue to evolve, taking a different shape and remaining necessary in practice.

It is my experience that the most effective way to implement digital construction is not by recruiting and forming a team of specialists, but by transforming existing roles within a company. While assembling a team of experts may be appealing, this approach is unsustainable and contradicts the goal of simplifying processes, enhancing productivity and reducing overhead costs. If a business does not support the upskilling of existing roles, it will struggle to internalise the necessary knowledge and cultivate the roots needed to integrate digital construction as a natural process.

Consequently, skills and capabilities will remain superficial, weak and fragile. If digital construction leaders or specialists leave the business, the company risks being unable to fulfil contractual obligations until a replacement is found. Therefore, incorporating the correct skills within existing roles is the most prudent approach to ensure a robust digital construction business and capabilities.

Developing a strategy to enhance existing employees' skills

Each business is unique, and roles may involve performing different tasks. It is fundamental to understand the activities conducted by various departments and identify their responsibilities within digital construction. Upskilling current employees, rather than relying on a large team of digital construction specialists, is a more efficient approach. Although a team of specialists may possess expertise and extensive technical knowledge and be able to comply with appropriate standards and best practices, they will not have the same comprehensive understanding of the project needs as the overall delivery team, which is involved in day-to-day operations and conversations. With the right digital construction knowledge, the delivery team is better equipped to make appropriate decisions based on project needs and avoid missing critical decision-making opportunities. Upskilling existing team members allows an organisation to build a strong knowledge foundation that can be leveraged to develop and implement digital construction strategies effectively. This approach not only reduces the risk of losing key personnel but also enables the organisation to adapt to changing circumstances and remain competitive in the market. Investing in upskilling existing team members is a wise and strategic decision that can yield a number of long-term benefits.

Embracing change in the construction industry

The construction industry has one of the lowest rates of productivity in comparison with other industries, largely because of outdated practices and resistance to new technologies. With the industry experiencing major disruption, now is the time for everyone to step out of their comfort zones, embrace change and undergo personal growth. This is not only about remaining competitive and relevant in the job market but also enhancing productivity and finding personal fulfilment. Adopting new technology and changing traditional approaches can be daunting, but it is a necessary step for staying relevant in a rapidly evolving world. By adapting to new technologies and transforming our roles, we can increase efficiency, reduce costs and improve the quality of our work.

Digital construction differentiates businesses from competitors in the short term and fosters long-term competitiveness by enhancing productivity, client satisfaction and ensuring access to accurate project information. This transformation requires unanimous support from all business levels, emphasising the importance of a well-crafted communication strategy. Introducing digital transformation can generate both enthusiasm and opposition. Strong support from senior leadership is also essential to overcome resistance to change.

Throughout my years of delivering training sessions, I have realised that age is not a decisive factor in adopting digital construction. People of all ages can embrace it; conversely, individuals of all ages can be less interested. Successful adoption relies on changing attitudes and behaviours rather than the technology or process itself. Young people often model their behaviour on those around them, making it all the more important for leaders to lead by example and initiate change from the top to drive the cultural shift.

Investing in training and support

To ensure a smooth transition, comprehensive training and support are crucial. You cannot expect a BIM specialist to deliver the project requirements single-handedly. It is better to draw on people in traditional roles, with the appropriate knowledge and skills, who are better equipped to make the best decisions for the project, considering each project's unique characteristics and the fast-paced nature of the construction industry.

When it comes to training, it is unwise to attempt to teach everyone everything. Instead, the content should be tailored to each role and focus on their specific goals. Not everyone needs or wants to know all the details regarding processes and technologies that do not directly impact their role. A knowledge base can always be provided, so teams can seek additional information if interested. Overloading individuals with too much information can quickly cause them to lose interest in the process, despite the promised benefits.

To ensure a successful training programme for digital construction transformation, follow these suggested steps.

- Plan the training thoroughly, considering the time required for each employee to complete it based on their role and responsibilities. Provide resources tailored to each person's needs.
- Keep the information readily available for consultation at any time, through an online portal or shared drive that is easily accessible. This allows employees to refer to the material whenever needed, reinforcing their learning.
- Regularly update the training resources, as the field of digital construction is constantly evolving. Without regular updates, employees may lose interest in the training material, feeling that the content is outdated.

Allocating time for training may be challenging for some organisations, but it is essential to secure the business's support in prioritising and investing in team development. By conveying the value and benefits of attending training sessions, you can encourage team members to actively participate and learn the necessary skills for implementing digital construction.

To create the right learning environment for success, employees should see that their managers and directors are fully engaged and supportive. This includes being willing to learn and implement training across the business, questioning teams about their completed training and ensuring that digital construction requirements are being met in projects. When employees see that their leaders are taking digital construction seriously, they are more likely to do the same. People tend to copy behaviours, so maintaining the right attitude and addressing those who do not support the business culture is crucial; otherwise, the change will not happen.

Culture of learning

By cultivating a culture of learning within the team, the business can experience numerous advantages that benefit both employees and the organisation.

- *Improved decision making.* With a comprehensive understanding of the process, teams can make well-informed decisions with various stakeholders at each project stage. This not only streamlines the decision making process but also enhances the quality of discussions, enabling team members to have appropriate conversations and even challenge some of the project requirements when necessary. Well-rounded knowledge reduces the need for consultation with a digital construction specialist, as team members themselves possess the correct understanding to guide the project effectively.
- *Career growth and retention.* The training programme supports employees' growth and learning, helping the business retain top talent by fostering a sense of belonging in a company

that values people's futures. Moreover, embracing technology can help attract and retain individuals who want to join the business, as such a culture can simplify tasks and enhance people's professional journeys. A culture of learning is crucial for business success.

- *Staying ahead of competition.* The training programme enables the business to stay up to date with the latest standards, services and technology in the market, providing better experiences for clients and maintaining a competitive edge.
- *Increased efficiency and profitability.* Enhancing efficiency and productivity in the workplace can lead to reduced overhead costs and increased profitability in the long term, positioning the business for greater success.

Maintaining accountability and consistency

While organisations typically empower leaders to innovate and develop their solutions, the adoption of digital construction necessitates a unified approach, and adhering to a digital construction strategy is crucial for success. In this scenario, the business needs employees who are willing to collaborate to enable the business to change and adapt to the new requirements. Additionally, it is essential that all team members are held accountable for fulfilling their BIM responsibilities within their respective roles.

Regular progress reviews, performance metrics and continuous improvement initiatives can help ensure that the organisation stays on track with its digital construction goals. By maintaining a strong commitment to accountability and consistency, the organisation can effectively implement digital construction strategies and achieve long-term success.

The objective of this strategy is to minimise dependency on a specific individual or team for the successful execution of digital construction projects. By disseminating knowledge and expertise among a larger number of team members, we aim to safeguard the project and the business as a whole. This approach ensures that if a crucial team member were to leave, the valuable knowhow they possess wouldn't be lost, thus preventing a detrimental impact on the entire business operation.

Taking inspiration from Winston S. Churchill (1961), 'To build may have to be the slow and laborious task of years. To destroy can be the thoughtless act of a single day.' Building a robust implementation of a digital construction approach within the business is a slow, meticulous process that demands exhaustive attention to detail and ongoing support. If the implementation is not done correctly, it can be quickly undone.

2.8. Engage your teams: why and what it means to them

The construction sector is undergoing a digital transformation, changing the way projects are planned, executed and managed. This is due to advances in processes and technology, as well as a growing emphasis on information management. As a result, many existing roles need to evolve to meet project requirements and remain relevant in the contemporary landscape. Your foremost duty, as a business leader, is to steer the change management process and support your team members as they adapt to new processes and project and business requirements.

It's essential for teams to understand that these changes are vital for the organisation's success and their individual progression. However, implementing change can be tricky; people frequently resist anything that upsets their established routines or challenges their beliefs.

To ensure a smooth transition, it is necessary that the team takes ownership of all initiatives. Each initiative should address the team's concerns and challenges in their day-to-day work to guarantee that the benefits of the solution are clear and wholeheartedly embraced. Without this clarity and ownership, it's easy to encounter resistance and pushback.

Promoting the benefits of digital construction to individuals

The changes needed for teams to embrace digital construction methods are small in comparison with the significant opportunities they provide. Nonetheless, teams often resist change until they see its positive impact on their daily work. As a digital construction leader, it is your responsibility to effectively communicate the benefits. This involves understanding what people want and helping them achieve it. Many may not realise what they can accomplish until they learn about the advantages that digital construction offers in their everyday roles.

Change within an organisation requires a joint effort and cannot be achieved by a single person. It is vital to have the support of the board of directors and senior leaders in the organisation. These individuals play a role in helping team members accept and embrace necessary changes for the organisation's growth and success. This includes holding them accountable, reminding them of the reasons behind the change and emphasising the benefits it will bring. To do this effectively, use empathy, provide necessary training and show support throughout the process while listening to feedback and concerns.

Everyone wants an easier work life, but it is your job to help people understand how digital construction can make that possible and create a demand for change.

To successfully adopt digital construction, clearly communicate and sell its benefits to all team members and involve them in the process. Begin by presenting a clear vision, highlighting the advantages and demonstrating the long-term value of digital construction.

Use your digital construction advocates to showcase the benefits and persuade teams to accept the changes. Understand your teams' needs and demonstrate how digital construction can enhance their day-to-day roles, fostering a desire for change. Engage with various departments, listen to different opinions and tailor your approach to address specific concerns and requirements. By emphasising the benefits and direct impact on each team member's work, leading to better project outcomes, you can effectively promote the advantages of digital construction.

Businesses often invest considerable effort in developing leaders, empowering them to innovate and devise solutions. However, there are times when these leaders need to follow a unified approach. In this case, it is crucial for the business to adhere to the agreed digital construction strategy. This should not be done in isolation but involve different individuals in the organisation, holding everyone accountable for fulfilling their BIM responsibilities as part of their roles.

When examining the roles within a main contracting business with key responsibilities in project delivery, it is important to offer clear examples of why it is necessary for the team to be on board, attend training sessions and embrace change. By effectively communicating the value and advantages of these efforts, you can ensure that your teams remain motivated and actively engaged in the process of adopting digital construction methods.

The following example demonstrates how to emphasise the distinct advantages to each role from adopting digital construction and how people's work will be enhanced. Remember that the specific benefits may differ based on your business, but this example illustrates the general concept.

Design manager and technical services manager

- Streamline coordination – saving precious time and boosting productivity.
- Mitigate risks during preconstruction – providing peace of mind and a smooth project flow.
- Enhance team collaboration, communication and information accuracy – reducing headaches and fostering a healthy work environment.
- Gain a deeper understanding of designs – optimising your time and improving your decision making.
- Strengthen communication and relationships with subcontractors – boosting job satisfaction and minimising hassles while enhancing project outcomes.

As a design manager or technical services manager, you'll save time, reduce risks and improve team dynamics. You'll make quicker, well-informed design decisions and foster better relationships with subcontractors. These benefits collectively enhance your effectiveness and position you as a key asset for successful project execution.

Preconstruction manager

- Deliver exceptional service to clients – enhancing job satisfaction, building strong client relationships and reducing hassles.
- Gain clarity on client BIM expectations – providing peace of mind, increased job satisfaction and a competitive edge in the market.

As a preconstruction manager, you're not only positioned to make your work life more gratifying but also to deepen relationships with your clients. You'll enjoy peace of mind, knowing you've clearly understood client expectations regarding BIM, thereby elevating your standing in the market. In short, you'll experience less hassle, greater job satisfaction and develop a competitive edge that sets you apart.

Commercial

- Accurately place subcontractor appointments – saving you time, money and frustration while avoiding costly mistakes.
- Assess subcontractor capabilities to deliver projects – giving you peace of mind and ensuring successful partnerships.
- Achieve better outcomes from subcontractors – fostering improved working relationships, cost savings and project success.
- Prevent additional costs and disputes during construction – reducing stress, saving time and money, and protecting your company's reputation.

As a commercial lead, you stand to gain not just in terms of time and cost savings, but also in the quality of your working relationships and project outcomes. Enjoy peace of mind, improved partnerships and reduced stress – factors that collectively contribute to safeguarding your company's reputation while elevating the success rate of your projects.

Estimator

- Quickly extract quantities from models – maximising efficiency and minimising errors.
- Enhance your understanding of designs using 3D models – granting peace of mind, job satisfaction and improved decision making capabilities.

As an estimator, you not only become more efficient and accurate in your work but also gain a deeper sense of professional satisfaction and empowerment in your decision making processes. This dual advantage places you in a stronger position in your role, potentially making you a more valuable asset to your team and organisation.

Planner
- Prevent delays caused by on-site rework and project reprogramming – saving valuable time and resources.
- Understand survey requirements to validate and verify the existing building conditions ensuring project accuracy from the start.

As a planner, you stand to gain both efficiency and reliability in your projects. This dual benefit not only enhances your value to the team but also significantly reduces risks associated with time and cost overruns, elevating the likelihood of project success.

Operations
- Ensure projects are delivered on time and within budget – optimising time and financial resources while exceeding clients' expectations.
- Minimise rework and defects – saving time and money, and building a solid reputation for quality work.
- Effectively communicate work and health and safety hazards – providing peace of mind, fostering a safe work environment and reducing liability risks.

In operations, you'll not only meet deadlines and budgets but also elevate client satisfaction. Your focus on quality minimises costly rework, while effective hazard communication ensures a safer, legally compliant workspace.

2.9. New responsibilities within existing roles

As mentioned in Section 2.7, to fully reap the benefits and ensure a successful implementation of digital construction, it is essential to develop the knowledge and skills of existing teams. This approach will prove more cost-effective and provide a solid foundation for the business, protecting its interests in the long run.

By investing in the training and development of the current workforce in digital construction, companies can increase profits, reduce overhead costs and allow team members to focus on delivering added value to clients. As the adoption of BIM in projects continues to grow and become the norm, it becomes increasingly critical for businesses to nurture their teams' expertise in this area.

During the transformation, it is crucial to assess existing roles within the organisation and determine which tasks related to digital construction, such as BIM, should be incorporated. The following responsibilities should be considered when embedding BIM within existing roles, keeping in mind that this mapping may not apply universally, as different companies have varying roles and structures. It is recommended to use this information as a reference and adapt it to your organisation's specific roles. A detailed breakdown of responsibilities for each role, at various stages of the information management process in accordance with ISO 19650-2 (BSI, 2021), is covered in Chapter 6.

It is also important to remember that these tasks should not replace or eliminate any other responsibilities that main contractors typically have in non-BIM projects. The implementation of digital construction should enhance and optimise current processes, not obstruct them. By fostering the growth of BIM knowledge and skills within existing teams, companies can establish a strong foundation and safeguard their business interests while maximising the potential of digital construction.

Before we delve into the specific adjustments and changes essential for each role, I'd like to share a significant message from Mark Shayler (2018). I had the opportunity to attend the BIM Awards in 2018, hosted by Shayler, an innovation and environmental consultant. One particular slide he presented had a lasting impact on me because it effectively highlighted some of the challenges the industry needs to address. The slide was so compelling that I took a photograph of it, and it contained the following messages, which I believe clearly represent the transformation that needs to occur at all levels within the industry:

"Disruption is normal.

Change is inevitable.

How it was done yesterday is not how it should be done today.

Nothing stays the same, and neither should you.

The problem arises when change happens and you don't."

Roles for a main contractor

Design manager and technical services manager

In a main contractor company, both the design manager and the technical services manager are vital for successfully incorporating BIM in the business and its projects. They each play a role in making sure that the design is not only accurate but also practical and safe, and that it meets all regulations. Given the overlap in their BIM duties, I've chosen to merge their responsibilities for this section. Generally, the technical services manager focuses on the design of services, while the design manager adopts a broader perspective. Both are instrumental in coordinating various task teams and assuring that the project aligns with its specified requirements.

TASK TEAM SELECTION

- Clarify the project's BIM expectations during the tender stage to make sure you get accurate quotes and the project's needs are met.
- Review the BIM assessment results that the digital construction lead provides. This helps evaluate the BIM capabilities of potential team members.
- Take into account notes and scores from previous performances on similar projects. Review historical data and client feedback for a comprehensive view.
- Stand firm during settlement discussions to make sure that team selection is based on capabilities, not solely on cost.

PROJECT RISK MANAGEMENT

- Keep the project risk register updated with any identified risks. Be sure to include input from task teams and the digital construction lead to help mitigate these risks.
- Keep the client in the loop by sharing risk information regularly.

- Use lessons learned from previous projects to improve delivery and minimise risks in the current project.

DETAILED RESPONSIBILITY MATRIX

- Provide the necessary input to create the BIM Execution Plan, and collaborate with the task teams to develop a detailed responsibility matrix that outlines roles and responsibilities across all teams.
- Establish key milestones for information delivery and create a timeline.

REVIEW OF TASK INFORMATION DELIVERY PLAN

- Make sure that the content of the proposed task information delivery plan (TIDP) matches the project's overall needs and programme.
- Review and comment on TIDPs before these are included in the appointment.

MASTER INFORMATION DELIVERY PLAN

- Keep the master information delivery plan (MIDP) up to date.
- Update the MIDP whenever there are changes in the TIDPs.

COMMON DATA ENVIRONMENT

- Define the workflows specific to the project for authorising and accepting information within the common data environment (CDE).
- Emphasise the value of collaboration and training for the entire team for the appropriate use of the CDE and information management.
- Ensure the project directory is up to date to reflect any changes in the project team.

COMMUNICATION AND SURVEY INFORMATION

- Collaborate with design disciplines and on-site teams to define the scope of the point cloud survey for accurate data capture.
- Train teams to make sure they understand the methods and procedures for information production.
- Check the survey information to make sure it's both accurate and suitable for developing or verifying the design.

DESIGN DEVELOPMENT AND COORDINATION

- Encourage open communication and collaborative problem-solving among task teams.
- Ensure prompt design resolution and collaboration. Coordinate efforts to resolve any design errors that may occur.
- Take charge of managing and verifying design changes to make sure they are accurately reflected.
- Advocate for the use of digital tools to improve design coordination and resolve clashes more efficiently.

QUALITY AND APPROVAL WORKFLOW

- Ensure a correct understanding of and verify compliance with the BIM Execution Plan by participating in discussions and addressing challenges.
- Take responsibility for reviewing, commenting, authorising or rejecting information at various quality assurance points.
- Ensure that the quality of information aligns with the TIDP and the BIM Execution Plan.
- Effectively manage the formal process for authorising and accepting project information.
- Verify the accuracy of COBie information provided by each discipline at different stages.

Information manager

The role of an information manager, for a main contractor, can vary significantly between organisations. In this context, the focus is on ensuring proper information management throughout the project. Information managers oversee the strategic handling of all data, can report and analyse the status of information within the common data environment (CDE) and are committed to meeting security requirements. While some may confuse the role of an information manager with that of a document controller, the latter is more specialised and concentrates solely on document management.

MANAGING NAMING CONVENTIONS AND PROTOCOLS

- Collaborate with the appointing party to establish a clear and standardised naming protocol, if one hasn't already been implemented for the project.
- Be certain that the protocol complies with BS EN ISO 19650 and is easy for all team members and stakeholders to understand.
- Keep the naming protocol updated, particularly when new companies join the project or new codes are introduced.
- Ensure that all project participants are well-informed about these updates for seamless information retrieval.

HANDLE THE COMMON DATA ENVIRONMENT

- Initiate the setup of the CDE during the tender phase and continue managing it until the project handover.
- Conduct tests on the CDE in the mobilisation phase to ensure that it functions properly and is accessible to all parties.
- Serve as the primary point of contact for any enquiries or issues related to the CDE. Provide targeted training sessions or tutorials to project teams to help them effectively use the CDE.
- Monitor and manage access permissions to the CDE, ensuring that only authorised personnel have access and revoking access for team members no longer involved in the project.

WORKFLOW MANAGEMENT AND TRAINING

- Collaborate with the design manager, technical services manager and client to establish and set up efficient project workflows, tailored to the project's requirements.
- Deliver thorough training sessions and provide instructional materials to help project team members effectively navigate and manage the common data environment.

QUALITY ASSURANCE AND REPORTING

- Perform the initial quality review of newly uploaded information to the CDE, ensuring that it meets the project's quality standards and is promptly shared with all relevant team members.
- Confirm that all uploaded information aligns with the TIDP in terms of naming protocol, metadata and delivery timings. Report any discrepancies to the delivery team.
- Produce weekly reports that outline pending actions within the CDE workflow. Share these reports with the delivery team to ensure project progress and promptly address any delays or issues.

OPERATIONS AND MAINTENANCE AND DATA COLLECTION

- Work closely with the site team to collect operation and maintenance (O&M) and asset data from various task teams. Make sure that the datasets align with project requirements and are accurately tagged with metadata.

ARCHIVING AND HANDOVER PROCEDURES

▨ Coordinate the transfer of information to the client's platform during handover, or at different project stages, in accordance with the client's requirements.

Preconstruction manager

The title of the role can differ, based on the company's size, structure and specific industry; other terms are bid manager and even senior estimator.

A preconstruction manager is crucial for the initial planning and setup of a construction project. The role goes beyond just winning the tender; preconstruction managers ensure that BIM requirements are considered and acted on from the outset. By choosing the right partners and clarifying requirements in a timely manner with both the client and task teams, preconstruction managers facilitate the successful implementation of BIM within the project.

VERIFY BIM DOCUMENTATION FOR TENDER

▨ Confirm that the client and its design team have supplied all necessary BIM documentation as part of the tender information.

DISSEMINATE DOCUMENTATION AND INFORM TEAM

▨ Distribute the received BIM documentation to the relevant parties within your organisation. Specifically, tell the digital construction lead about the project details and extend an invitation to the tender launch meeting.

ADVOCATE FOR BIM ADOPTION

▨ If the client has not yet considered adopting BIM, help them understand the benefits of BIM at various project stages. Encourage the client to embrace the BIM process.

▨ Emphasise the importance of implementing BIM during the early stages of conception and design for maximum effectiveness.

HANDLE INVITATION TO TENDER (ITT) QUERIES

▨ Work closely with the digital construction lead to handle any specific BIM-related questions that may come up during the ITT process.

▨ Compile and present the required evidence and case studies to meet the BIM-related criteria outlined in the ITT.

TASK TEAM SELECTION AND COMPLIANCE

▨ Ensure the selection process for task teams includes a thorough evaluation of their BIM capabilities, to confirm they are adequately prepared to fulfil the project's BIM requirements.

▨ Safeguard the BIM requirements from being altered or removed as the project progresses, ensuring that the project remains in compliance with company policies.

MANAGE CONTRACTUAL DOCUMENTATION

▨ Understand the contractual documents, such as the information standard and protocol, appointing party's exchange information requirements (EIR), BIM Execution Plan and master information delivery plan (MIDP).

▨ Ensure that the relevant documents are included in the appointment.

▨ Implement proper change control mechanisms to document and manage any modifications to these key documents.

Estimators and commercial team

The estimators and commercial team, or quantity surveyors, depending on the company, are essential for conveying BIM requirements. Both roles involve ensuring that task teams have the precise information needed to understand the project scope and provide accurate cost estimates. Although the responsibilities during the tender process are similar, tasks related to appointment and delivery and account closure fall under the commercial role.

TENDER PROCESS

- Ensure that all task teams, with design and data requirements, complete the required BIM assessment to determine their suitability for the project. This assessment should be submitted as part of their tender return.
- Distribute BIM documentation to potential partners and clarify the project's BIM expectations to ensure that quotations are accurate.
- Make sure that all parties have the correct documentation, including the ability to access any existing 3D models, allowing them to review the information for a more accurate quotation.
- Exercise caution when sharing sensitive information and adhere to security protocols when distributing BIM documentation or providing model access.
- Question any additional costs that task teams might propose for fulfilling BIM requirements. Evaluate the financial implications of point cloud surveys and BIM-related fees.
- Utilise digital tools for a comprehensive understanding of the project's scope and design, as well as for quantity extraction if applicable.
- Ensure that the task information delivery plan (TIDP) is completed as part of the tender return as it is required for the appointment process.
- Understand the BIM expectations for each task team and provide them with appropriate information based on their scope, avoiding overwhelming them with irrelevant details.

APPOINTMENTS

- Ensure a comprehensive understanding of the level of BIM compliance and involvement required from each task team during the project.
- Incorporate relevant BIM-related documents in all appointments, such as project information standards, exchange information requirements, BIM Execution Plan and TIDP.

PROJECT DELIVERY AND ACCOUNT CLOSURE

- Before closing any accounts, ensure that all parties have fulfilled their BIM obligations as outlined in their respective appointments. This includes delivering the correct design models and asset data.
- Additionally, support the digital construction lead and information manager in obtaining appropriate and accurate information from task teams.

Planner

The planner is tasked with formulating, managing and updating the project's schedules.

PRECONSTRUCTION ACTIVITIES

- Consider the time implications of conducting the point cloud survey as part of the critical path and determine the best time to carry out this activity.
- Collaborate with the design manager and technical services manager to ensure that the MIDP meets the programme requirements.

- Understand the impact of any delays in processing information on the programme and take lessons learned from previous projects into consideration.

- Utilise digital tools and inspect the model to gain a comprehensive understanding of the project for programme creation.

- Support the production of visualisations that illustrate logistics and the work to be completed as part of the project.

Operations

The operations team might include the building manager or construction manager, depending on the business. These individuals are closely engaged with task teams during the project's execution stage, making them crucial players in ensuring the successful implementation of BIM. By providing support to different task teams and verifying that work is carried out as planned, they play an important role in effectively adopting BIM during the construction phase and ensuring the correct handover of information to the client.

FOSTER A COLLABORATIVE AND LEARNING ENVIRONMENT

- Establish a collaborative digital construction atmosphere, where all task team members attend training sessions provided by the digital construction lead and the information manager.

- Encourage the use of digital tools for real-time model access and data review, discouraging the use of printed drawings to prevent the risk of using outdated or superseded project information.

- Ensure that the master information delivery plan (MIDP) is actively updated to address any delays or issues with task teams.

- Utilise reports and feedback from the information manager to enhance workflows and communication, ensuring that the task team takes action to maintain smooth and up-to-date information exchange.

- Conduct regular on-site model reviews involving all key stakeholders and ensure that the team uses the model to address design challenges and enhance collaboration and coordination.

- Ensure that the work matches the agreed design, and any agreed changes are reflected in the design information, including the model, before execution.

DATA VERIFICATION AND CLIENT ENGAGEMENT

- Collaborate with the design manager and technical services manager to verify the accuracy of the task team's submitted asset data. This verification ensures that the data accurately reflect the actual progress and quality of work on site.

- Ensure that all project data and design information are promptly updated to incorporate any modifications agreed on during on-site work. This supports the client's needs during the asset's operations and maintenance phase.

- Verify and accept O&M information to confirm accuracy and compliance of actual work conducted on site with client requirements.

- Maintain close collaboration with the client throughout the project and explain the use of BIM during the project's various phases.

- Work closely with the digital construction lead to identify areas for future success and improvement. This includes considering feedback from clients, the internal team and task teams.

Digital construction lead

The role of digital construction lead can vary, depending on the company's size and strategy. However, in this context, the primary function of the digital construction lead is to support teams in accomplishing their main tasks and to oversee the implementation of BIM within the project.

INFORMATION ASSESSMENT AT TENDER STAGE

- Attend the tender launch to gain a comprehensive understanding of project requirements, scope and potential challenges.

- Thoroughly assess all tender documentation to pinpoint both opportunities and risks and communicate these findings to the preconstruction team.

- Raise queries for any unclear or ambiguous specifications in the project tender. Engage in direct conversations with the appointing party during the tender phase to address missing information and challenge assumptions that could hinder the project's success.

- Review the BIM assessment responses for each task team and provide feedback to the preconstruction team regarding the results to assist in selecting appropriate teams.

- Help the client understand the BIM process by presenting its benefits and supporting your points with case studies from previous projects.

- If the design has already been produced and the BIM Execution Plan is available, carefully examine the quality of the project deliverables in comparison with the agreed plan and the client's exchange information requirements (EIR). Make sure they match and are consistent.

- Also, take into account the impact of the contractor's design portion (CDP) and set out the necessary compliance requirements for the task teams.

PLANNING AND DOCUMENTATION

- Spot and articulate any BIM-related risks that could delay project delivery and ensure these are logged in the risk register.

- Work closely with the project team to formulate a detailed mobilisation plan that outlines the procedures, training needs and digital tools required.

- Settle on agreed methods for information production, naming conventions and survey strategies.

- Create a pre-appointment BIM Execution Plan based on the exchange information requirements (EIR) from the appointing party and input from known task teams as part of the tender response. Update it accordingly if the tender is successful.

TRAINING AND SKILL DEVELOPMENT

- Arrange mandatory training sessions that cover the project scope and BIM processes, incorporating lessons learned from previous projects.

- Ensure that all team members, both internal and external, have the necessary skills and training to meet BIM requirements.

- Establish and grant access to all digital tools required for project delivery and provide training on their usage.

- Supply the team with necessary resources, such as file format guidelines and survey tools, to enable efficient workflow.

MONITORING AND PROGRESS TRACKING

■ Produce regular reports that track the progress and quality of the information provided at each project stage and validate the accuracy of the information.

■ Record the performance metrics of task teams to assist in future team selection and inform business decision making.

■ Bridge digital gaps. Be accessible to various teams to eliminate barriers between digital construction processes and operational teams, thus fostering successful BIM implementation.

■ Continue ongoing conversations with the team to identify challenges and promptly address any non-compliance issues with contractual requirements.

FEEDBACK LOOP AND FUTURE PLANNING

■ Record stakeholder feedback and lessons learned at various project stages to benefit future projects.

■ Create and share marketing content that showcases project achievements and identifies areas for improvement for both internal and external communication.

■ Gather data for analytics to assist in making informed business decisions and update company documents using insights gained from past projects.

Roles for a design practice

The process of implementing digital construction may present similar challenges for subcontractors, design consultancy firms and main contractors, but the specific responsibilities and roles involved will vary. However, the ultimate goal remains the same for all: to enhance the skills of existing roles for better design and information management.

The following approach is an example of how this might be executed in an architecture firm. It is not meant to be a one-size-fits-all solution, but rather a starting point for companies to assess and adapt to their specific needs.

For successful digital construction implementation, each company must evaluate its unique roles, responsibilities and goals. Doing this will enable you to effectively incorporate digital construction into your existing processes and ensure you can achieve the ultimate goal of improved design and information management.

Designers

The role of a designer can vary, depending on the company and trade, but generally, designers are responsible for generating information and contributing to the overall design process.

DESIGN AND MODELLING BEST PRACTICES

■ Follow industry-standard best practices to achieve high-quality design outcomes and accurate modelling production.

■ Actively participate in initial kick-off meetings and subsequent design development meetings with representatives from other disciplines involved in the project.

BIM EXECUTION PLAN AND STANDARDS COMPLIANCE

■ Rigorously adhere to the guidelines, standards and procedures outlined in the BIM Execution Plan.

▨ Conduct checks against information standards, recording the outcomes for quality assurance and compliance monitoring.

PROJECT MANAGEMENT AND INFORMATION SUBMISSION

▨ Take charge of various aspects, such as design, model maintenance, coordination, project organisation, document revisions, archiving and publishing.

▨ Periodically update the project's information model based on the current project status and any changes that have occurred.

▨ Submit project information and associated metadata punctually to the common data environment for review and approval.

CHANGE CONTROL AND REVISION

▨ Revise and update design information in accordance with any new discoveries or changes that occur during the construction process. All changes must adhere to the appropriate change control processes.

Design managers

The name of the role can vary, but within a task team the design manager generally coordinates the entire design process and serves as the point of contact for clients and task teams, making sure that the design meets the project's needs.

BIM ASSESSMENT AND BIM EXECUTION PLAN REVIEW

▨ Complete the BIM assessment in an honest manner and provide constructive feedback to the lead appointed party.

▨ Attend BIM launch meetings to familiarise yourself with the project's scope and review the BIM Execution Plan (BEP) for compliance and alignment.

▨ Thoroughly understand the project BIM requirements and the overall scope of the project.

▨ Plan and schedule all design deliverables and ensure the timely completion of the task information delivery plan (TIDP).

DESIGN COORDINATION AND ISSUE RESOLUTION

▨ Oversee the coordination of all design-related activities within the project, ensuring that any issues are promptly identified and resolved.

INFORMATION REVIEW AND GENERATION

▨ Regularly review and provide comments on the project information stored in the common data environment. Ensure that these comments are subsequently incorporated in the project's design.

▨ Generate accurate project information in accordance with the agreed TIDP and project information standards, and approve or reject information containers after internal quality checks.

COORDINATION MEETINGS

▨ Attend regular coordination meetings to discuss and resolve any design, process or compliance issues.

Digital construction lead

Within a task team, the digital construction lead oversees the use of digital technologies, as well as training staff, developing digital strategies and ensuring efficient workflows. The digital construction lead is primarily responsible for improving efficiency and driving innovation within the organisation.

BIM EXECUTION PLAN AND TRAINING

▦ Conduct a thorough review of the project BIM Execution Plan and share any comments, concerns or suggestions with the lead appointed party.

▦ Produce the BIM Execution Plan, if it's not available, based on the project requirements and the client's EIR.

▦ Provide necessary BIM training sessions to team members, identifying areas that need skill improvement.

STANDARDS AND PROCESS MONITORING

▦ Monitor and ensure effective implementation of BIM standards and procedures in the project.

DASHBOARD MANAGEMENT AND DATA ANALYSIS

▦ Oversee the creation and maintenance of BIM process dashboards, using them to identify and investigate any data anomalies that could delay the project.

DIGITAL TOOLS AND PROJECT INFORMATION

▦ Ensure that the team makes appropriate and efficient use of digital design tools available, always following best practices for model production.

▦ Verify that all project information is delivered in accordance with the standards and procedures established in the BEP.

PROJECT SETUP AND ASSET MANAGEMENT

▦ Support project setups by providing correct templates that align with the company's standards.

▦ Make sure that the family libraries used in models are kept up to date, are modelled efficiently and are not resource-intensive.

KNOWLEDGE SHARING AND COORDINATION

▦ Capture valuable lessons learned during the project and disseminate this knowledge within the organisation for continuous improvement.

▦ Coordinate the creation and sharing of geometric models among various teams and resolve or escalate any issues to the lead appointed party for final resolution.

BIBLIOGRAPHY

BSI (2013) PAS 1192-2:2013: Specification for information management for the capital/delivery phase of construction projects using building information modelling. BSI, London, UK.

BSI (2019) BS EN ISO 19650-1:2018: Organization and digitization of information about buildings and civil engineering works, including building information modelling (BIM). Information management using building information modelling. Part 1: Concepts and principles. BSI, London, UK.

BSI (2020a) BS EN ISO 19650-3:2020: Organization and digitization of information about buildings and civil engineering works, including building information modelling (BIM). Information management using building information modelling. Part 3: Operational phase of the assets. BSI, London, UK.

BSI (2020b) BS EN ISO 19650-5:2020: Organization and digitization of information about buildings and civil engineering works, including building information modelling (BIM). Information management using building information modelling. Part 5: Security-minded approach to information management. BSI, London, UK.

BSI (2021) BS EN ISO 19650-2:2018 & Revised NA: Organization and digitization of information about buildings and civil engineering works, including building information modelling (BIM).

Information management using building information modelling. Part 2: Delivery phase of the assets. BSI, London, UK.

BSI (2022) BS EN ISO 19650-4:2022: Organization and digitization of information about buildings and civil engineering works, including building information modelling (BIM). Information management using building information modelling. Part 4: Information exchange. BSI, London, UK.

Cabinet Office (2011) *Government Construction Strategy*. Cabinet Office, London, UK. https://assets.publishing.service.gov.uk/media/5a78ce8eed915d07d35b2933/Government-Construction-Strategy_0.pdf (accessed 23/11/2023).

Cambridge Dictionary (2023) FOMO. https://dictionary.cambridge.org/dictionary/english/fomo (accessed 17/11/2023).

Churchill WS (1961) *The Unwritten Alliance: Speeches 1953–1959*. Cassell, London, UK.

HMG (Her Majesty's Government) (2022) Building Safety Act 2022. The Stationery Office, London, UK.

IPA (Infrastructure and Projects Authority) (2021) Transforming Infrastructure Performance: Roadmap to 2030. https://www.gov.uk/government/publications/transforming-infrastructure-performance-roadmap-to-2030/transforming-infrastructure-performance-roadmap-to-2030 (accessed 23/11/2023).

Maxwell JC (2001) *The Power of Leadership*. Honor Books, Colorado Springs, CO, USA.

Shayler M (2018) *Presentation at BIM Awards*. BIM Show Live, Newcastle upon Tyne, UK. See https://www.bimplus.co.uk/willmott-dixon-interiors-and-david-philp-lead-winn/ for further details (accessed 23/11/2023).

Amador Caballero
ISBN 978-1-83549-446-2
https://doi.org/10.1680/iceedc.9446203

Chapter 3
Supply chain

3.1.　Introduction

The success of any project is tightly bound to the supply chain that supports it. As the well-known adage says, 'You're only as strong as your weakest link.' In this chapter on supply chain management, I'll discuss the importance of selecting the right business partners, understanding their capabilities and nurturing a collaborative relationship to ensure that your projects succeed.

Choosing the wrong task team can significantly increase the number of hassles and the cost of error during project delivery. Therefore, it's vital to stress the importance of making well-informed decisions for successful project delivery.

This chapter will explore the nuances of selecting task teams, emphasising the need to interview potential task teams to understand their culture, as this underlines the importance of assessing them before bringing them on board for new projects. This approach will offer more insight than any questionnaire you send them to complete. It's crucial to evaluate not just their technical skills but also their values and vision. Partnering with a company that aligns with your digital goals can lay a strong foundation for future growth and success. Additionally, consider the performance of the supply chain in previous projects, learn from past challenges and use this information when selecting your partners for future projects.

When deciding on a task team, cost should not be the only criterion. The cheapest option often yields poor service, while the most expensive does not necessarily guarantee superior performance. A balanced understanding of the supply chain, including the team's capabilities, capacity to meet project requirements and culture, is essential. While initial assessments are important for making a selection, good performance during the assessment phase does not guarantee future success in executing the work. Therefore, it's crucial to continuously monitor the performance throughout the project, taking note of both positive and negative experiences for future reference.

Understanding your supply chain's approach to digital construction, along with the benefits and challenges faced during design and construction and the support that the team needs from you, is required. By aligning with partners who share your digital vision, you create a cooperative, innovative and successful working environment.

It is important that you support your trusted partners in embracing digital transformation and work with them to achieve the level of skills and capabilities that you need from them. Offering the right support and resources can help them adapt to the rapidly evolving industry landscape, thus ensuring that your combined efforts produce the best possible outcomes.

Knowing the various task teams involved in a project and understanding their responsibilities in the digital construction process is vital for effective management during the tender stage. Providing the right documents without overwhelming your task teams helps maintain an efficient working

55

relationship. This enables you to fine-tune your approach and communication, leading to more successful tenders and project results.

In conclusion, carefully selecting and supporting your partners will not only ensure the success of your projects but also keep you competitive in an increasingly digital landscape. This chapter will guide you through the process of building and nurturing these relationships, providing you with the tools and insights necessary to achieve project success. With a focus on understanding your supply chain culture, assessing their capabilities and offering support to your partners, you will be well-equipped to navigate the complexities of the digital construction landscape.

3.2.   Adopting digital construction to remain competitive

To remain competitive, adopting digital construction and integrating the BIM process are essential steps. These not only improve project delivery but also enhance the management of information more efficiently, thereby increasing productivity by saving time and minimising errors. Contrary to what some companies believe, properly integrating BIM into a project can lead to a reduction of costs, enhancing competitive advantage.

However, over the years, I have seen how some companies inflate their fees when they see the term 'BIM' in documentation. This is akin to the way costs can suddenly multiply when the word 'wedding' is mentioned in other contexts. While there are legitimate scenarios where fees may increase owing to specialised client needs or requirements, such as weekly point cloud surveys or excessive asset data needs, raising fees should not be standard practice if the company has the correct capabilities.

Selecting the right supply chain has always been a crucial aspect of successful delivery; in the current transition to a digital environment, it becomes even more essential. Choosing the right partners requires attention not only to their technical expertise but also to their alignment with the culture and philosophy of your business. A team that is wholly committed to the digital construction approach can spell the difference between success and failure for a project.

Some companies might not feel the urgency to embrace a digital approach, especially if their clients have not asked for it. Others might lack the resources or fear the internal changes required. However, the cost of transitioning to digital construction isn't solely related to hardware; it involves a strategic investment of time to analyse, standardise and simplify current processes. These efforts can yield significant gains in the medium to long term.

As the construction industry evolves, clients are increasingly expecting precise information management and accuracy. Falling behind in the digital transformation that meets these expectations is not an option for companies aiming to compete and win new contracts. It's essential for businesses to proactively adopt innovative methods and demonstrate excellence in information management. This commitment safeguards operations and ensures compliance with vital legal standards, such as the Building Safety Act 2022 (HMG, 2022). But it's more than just adhering to regulations; it's about positioning the company at the forefront of a rapidly advancing industry, ready to meet the demands of today and tomorrow.

In my opinion, companies that resist change and continue to inflate fees to meet client needs must recognise that transformation is not optional. Embracing digital construction and BIM is a strategic necessity to maintain competitiveness. As rivals move towards adopting this transformation, standing still is no longer an option. Digital construction represents a fundamental shift in the industry, providing tangible benefits in efficiency, accuracy and competitiveness. It's a fundamental step for any company aiming to thrive in the construction landscape of the future.

3.3.   Interview and know your supply chain culture

The implementation of digital construction presents significant challenges, despite the numerous benefits it brings to businesses. To ensure success, I recommend adoption even when clients might not be interested, as it fosters a natural working culture within the company and enhances the skills of the team.

However, in my experience, while the businesses I have worked with have generally recognised the benefits of this approach, some commercial team members have highlighted the potential negative impact on our commercial competitiveness of implementing the BIM process in projects where the client had no interest, and when our competitors did not have the same approach. This misalignment in methodologies could give our competitors an advantage, potentially leading to loss of business opportunities.

In that case, the concerns of the commercial team were fair, and it was due to such concerns that the preconstruction team with whom I was working noticed that, in some instances, the fees quoted by the supply chain were significantly higher when the BIM requirements were included as part of the tender enquiry.

Choosing the right task teams for the project is crucial to maintaining a competitive edge and ensuring smooth project delivery. It is essential to assess companies based on their capabilities and capacity, as well as previous performance in delivering digital construction. Most importantly, it is necessary to understand the company culture and approach to digital construction; having a preference for collaboration and a solution-oriented mindset over rigid contract negotiations is necessary for a smooth project delivery.

While it may be tempting to select partners based solely on cost, it is essential to look beyond the initial financial aspect. The lowest bidder might not necessarily be the most competent in digital construction practices, and choosing them could lead to potential issues during project execution.

While it isn't necessary to cease collaboration with specific companies with whom you have a great business relationship because they lack the appropriate digital construction capabilities, you do have a responsibility to assist them in their journey to adopt digital construction. Helping them evolve into the partner your business requires is crucial. Otherwise, you might need to reassess your current supply chain's ability to support you with the delivery of digital construction projects. Collaborating with a company lacking the needed qualities could lead to issues and delays during project delivery.

I firmly believe that the adoption of BIM significantly enhances the efficiency of project delivery. So when, on one occasion, I noticed that certain members of the supply chain were raising their fees as soon as 'BIM' was mentioned in project documentation, I felt an urgent need to understand the specific reasons behind these fee increases. More importantly, I wanted a comprehensive insight into these companies' capabilities, approach and expertise, to ensure that they were suitably equipped to help us effectively execute and deliver digital construction projects.

As mentioned, in some instances of increased fees may be justified when clients have unrealistic expectations. In such cases, additional resources may be necessary, leading to an increase in cost. However, on projects where BIM is implemented from the start with reasonable expectations, it should not be the case that an increase in fees is requested.

To address these concerns, I conducted personal interviews with key companies to understand their approach to BIM and why they increased fees for BIM projects. The interviews helped me gain insight into the culture and approach of the supply chain we were considering working with.

During these interviews, it became apparent that the increase in fees was not directly related to the BIM process itself but rather to the limited skills and experience of the teams involved. Many task teams had already implemented BIM as a standard practice without additional fees. However, some companies charged higher fees for projects that required structured asset data, such as COBie data delivery, based on client expectations, even if this was basic information. Others did not have an in-house team to deliver these requirements, necessitating subcontracting the BIM deliverables.

With this in mind, I would like to share the findings from my discussions with the supply chain. I placed particular emphasis on mechanical and electrical subcontractors, given their highly influential role in both the design and construction phases of projects.

Common benefits of implementing BIM

During the interviews, all the teams recognised the numerous benefits associated with implementing BIM throughout the design, construction and operations and maintenance phases of a project. It was unanimously agreed that BIM should be implemented from the project's outset to maximise these benefits. Here are the key benefits identified by the teams.

Improved collaboration and information management

The BIM process facilitates enhanced collaboration and information management. A common data environment prevents information loss and minimises potential disputes. With access to the same information, team members can work together more effectively, fostering a collaborative environment and reducing misunderstandings.

Enhanced coordination and efficiency

One of the most significant advantages of BIM is the ability to work with a 3D digital model of the project. This promotes better coordination between team members and improves overall efficiency. Utilisation of the 3D model can lead to faster installations and fewer errors, ultimately increasing productivity throughout the project.

Streamlined prefabrication and manufacturing process

The BIM process streamlines the prefabrication and manufacturing process by expediting on-site work and minimising waste. Components are manufactured in a controlled environment before they are transported to the site for assembly. This approach enhances the quality of the components, reduces waste, improves efficiency and shortens construction timelines. Additionally, improved design coordination allows manufacturers to produce items with greater accuracy, minimising on-site corrections and waste. This increased efficiency saves both time and money, resulting in a more successful project outcome.

Improved procurement planning

In BIM, clear and comprehensive information is provided about the project's logistics and required quantities, which greatly aids the procurement process. This clarity enables task teams to plan their work more effectively, ensuring timely and accurate procurement of materials and resources.

Reduced issues and enhanced client satisfaction

By utilising BIM, the number of issues that arise during the project can be significantly reduced; BIM also enables the provision of better information to clients on project handover. These benefits result in the delivery of superior quality projects and ultimately enhance client satisfaction.

The common agreement among all the interviewed teams regarding the numerous advantages of BIM, especially when implemented correctly from the project's inception, was truly a breath of fresh air.

Common challenges in implementing BIM

While acknowledging the numerous advantages of implementing BIM, the teams did express some reservations about the common challenges they faced. Addressing these challenges is vital to ensure a more successful application of BIM in projects.

Insufficient information and coordination

One of the key challenges identified by the teams revolves around the issue of insufficient coordination and detail in the information provided by mechanical and electrical (M&E) consultants during the tender stage. While the M&E design consultants recognise the benefits of involving an M&E contractor during the design development phase to improve the quality of the design and ensure its buildability, it is evident that this collaboration is not always possible; in many cases, the information provided to the M&E consultants lacks the necessary coordination and level of detail required for seamless project execution.

Subcontractors often find themselves facing excessive caveats and scope gaps in the information provided by M&E consultants, which leads to additional efforts and costs required to de-risk the project. The frustration caused by these shortcomings can result in delays in the project timeline and increased fees, as the subcontractors need to spend additional time and resources to mitigate the risks and ambiguities in the design.

Addressing this challenge requires the establishment of a clear understanding of deliverables and responsibilities between M&E consultants and subcontractors during RIBA Stages 3 and 4. It is especially important to emphasise this during RIBA Stage 4, which may be referred to as Stages 4a, 4b and 4c in some projects. Therefore, at the tender stage, it is essential to provide all necessary information and clarify project requirements comprehensively. This ensures that subcontractors can accurately price the project, minimising the need for assumptions and reducing associated risks.

Need for design adjustments

Another challenge that arises is when the M&E consultant's design fails to meet the contractor's expectations. In such cases, subcontractors may be compelled to develop their own design from scratch, which can be a time-consuming process. As subcontractors often face pressure to start work promptly, there is a critical need for sufficient time to review the tender stage information, provide feedback to the design team and main contractor, and finalise design coordination before initiating on-site activities.

By allowing adequate time for design coordination and adjustment, subcontractors can ensure that the design aligns with their construction methodologies, sequencing requirements and procurement strategies. This proactive approach helps in avoiding potential conflicts and delays during the construction phase, as the design is thoroughly reviewed and adjusted to meet the contractor's specific needs and preferences. Early involvement and collaboration between the M&E design consultant and the subcontractor play a vital role in minimising design adjustments and facilitating a smoother transition from design to construction.

Enhancing project efficiency and quality
Addressing these challenges is of paramount importance to improve project delivery efficiency and quality. Recognising the benefits of involving an M&E contractor during the design development phase is crucial in optimising the design and ensuring its feasibility and constructability. By fostering effective communication, collaboration and coordination between all stakeholders, the project can benefit from their collective expertise and insights.

By providing comprehensive and well-coordinated information during the tender stage, the likelihood of delays and disruptions can be significantly reduced. Subcontractors can accurately price the project, minimising assumptions and risks associated with inadequate information. Allocating sufficient time during the procurement stage for such activities as reviewing information, providing feedback and finalising design coordination is essential and should not be overlooked in the tender programme. A collaborative approach that emphasises the integration of design and construction considerations can lead to improved project outcomes, with optimised designs, minimised conflicts, reduced rework and enhanced project efficiency and quality.

Support requested to implement BIM
Considering the challenges mentioned earlier, to effectively deploy BIM and enhance the efficiency and success of a construction project, specific support and considerations are essential for engaging correctly with the supply chain. Here are the key areas highlighted that require support.

Sufficient time for procurement and design review
Task teams need adequate time during the procurement stage to thoroughly review project specifications and complete a coordinated design before commencing work on site. This time allowance ensures that task teams can fully understand the project requirements and assess the feasibility and constructability of the design. With enough time, task teams can enhance the prefabrication process, making on-site operations more efficient. Therefore, the programme should account for this added time, allowing the contractor to comprehensively review information and make the necessary changes in the design if needed, before initiating the work.

Clear definition of project stages and deliverables
The smooth progression of a project relies on the definition of clear and specific deliverables for each stage, fostering mutual understanding among all project members. It's important to recognise the potential for confusion, owing to differing industry standards. Many project management guidelines, like those from the Royal Institute of British Architects (RIBA) and the Building Services Research and Information Association (BSRIA), offer varying project stage definitions, which might result in misunderstandings. Moreover, it's crucial to communicate openly with task teams about any unavailable anticipated deliverables for a given stage and ask for their further development, ensuring consistent progress within the set project framework. Promptly notifying missing deliverables is the key to preventing potential issues.

The role of an on-site leader

A dedicated on-site leader is essential for the seamless progression of a project. While this person doesn't need to be a digital construction specialist, the design manager, technical services manager or building manager will often take this role. These people ensure that task teams collaboratively address design issues and resolve conflicts rapidly, keeping the project on track. Furthermore, they prioritise timely information sharing with all stakeholders for transparency and adherence to the schedule. Without the right leadership, unresolved design issues and a lack of collaboration between different teams will hinder project delivery.

Comprehensive tender information

Providing task teams with the necessary information is crucial for the successful implementation of BIM. This encompasses not just the BIM Execution Plan, complete with detailed asset information requirements (AIR) in the tender documents, but also other available resources. These resources might include surveys, models or asset data, among other examples. They allow the task teams to review and evaluate the quality and level of coordination within the provided information during the tender stage. Active communication is crucial at this stage to address and alleviate any concerns.

Moreover, to expedite and continue the design by the subcontractors, it's essential that consultants share their models, retaining views, sheets and tags, preferably in the native file format, especially when teams are using the same authoring tool. Doing so accelerates the process, eliminating the need to replicate work already done by the design consultant. If this information isn't accessible, it's imperative to communicate this fact, as additional time will be needed by the subcontractors for this task.

These steps ensure that task teams grasp the project requirements thoroughly and can plan effective strategies.

Early contractor engagement

Engaging subcontractors early on, ideally around RIBA Stage 3, to oversee the development of the M&E design and assist consultants in finalising RIBA Stage 4, is a strategic response to the challenges and needs previously outlined. This early involvement enables subcontractors to provide input during the design phase, ensuring a smoother transition to the procurement phase. Through active collaboration with design consultants, subcontractors gain a more profound understanding of the project's requirements, thereby streamlining the procurement process. Establishing this partnership with your trusted supply chain leads to key advantages, such as a more solid design at an earlier stage, increased familiarity with the project and enhanced execution.

Conclusions

The following summary presents my personal conclusions, derived from extensive interviews conducted with different task teams.

Capabilities

One of the pivotal aspects that emerged from my interviews with different task teams is the necessity for task teams to have in-house BIM capabilities. These capabilities are essential in fulfilling project requirements efficiently and effectively. The BIM process involves managing accurate information and developing designs within a 3D digital environment. The vast majority of the task teams interviewed displayed a comprehensive understanding of the importance of this service. They had either already built these capabilities in-house or were in the process of doing so. However, some task teams were still in the learning phase, primarily owing to a lack of substantial exposure to projects with BIM-specific requirements.

All task teams, without exception, recognised the inherent advantages of providing structured data and accurate information to support the client during an asset's operations and maintenance phase. This understanding extended to the realisation of how such data can support the main contractor in tracking and tracing installed products, which in turn allows for any issues or concerns to be addressed swiftly and efficiently.

The issue of meeting basic COBie requirements surfaced during the interviews, revealing a spectrum of responses. While many task teams were willing to fulfil these requirements without charging an extra fee, others required an additional fee to deliver basic COBie data.

This is especially important if COBie is not a mandated requirement by the client or by your company's policy. By meticulously evaluating the capabilities and capacity of potential task teams and the project requirements, one can choose a partner that delivers high-quality results within the stipulated time frame and to the required standards.

Cost

The cost of implementing BIM was a significant concern that initially prompted me to conduct interviews with M&E subcontractors. However, the responses received did not, in general, reflect this concern. Most task teams interviewed did not include additional fees for BIM in their tenders, provided that the BIM requirements were clearly communicated and implemented from the beginning of the project. They did, however, acknowledge that an extra fee would be necessary for projects with extensive COBie requirements. This could become a point of concern in applying a policy to provide BIM on all projects, as the additional cost to deliver the client's COBie requirements would apply to all main contractors bidding for the same project.

If an M&E contractor includes a cost in the tender, it is crucial to understand the reasoning behind this cost. All task teams interviewed recognised the inherent benefits of utilising BIM in construction projects. They acknowledged its potential to help them deliver projects more efficiently and more effectively. Thus, it is vital to consider the potential costs proposed by task teams and the benefits of using BIM when evaluating task teams for a project.

Management

The interviews with subcontractors reinforced a crucial point we have already emphasised: the importance of engaging them as early as possible in the project. This early engagement, coupled with the provision of adequate time to review the work completed by consultants and finalise the design, if necessary, allows subcontractors to better understand a project's requirements. This understanding ensures that subcontractors have the necessary information to accurately price the project and plan their work accordingly. The subcontractors I interviewed also suggested that collaborating with design consultants to improve work at RIBA Stages 3 and 4 would be beneficial for all parties involved. Such collaboration would enable a more feasible and realistic design.

One of the essential takeaways from these interviews was the imperative need for transparent communication about the project status and the expectations set for subcontractors. These stipulations should be informed by the work that consultants have completed in the earlier stages and the deliverables that are available, based on the current phase of the project.

Another vital consideration is the requirement for additional time to review and interpret information. This extra time is key to mitigating additional charges to cover potential risks. Furthermore, delivering comprehensive and precise information at the outset can significantly reduce the scope for assumptions. This clarity enhances the certainty surrounding the project's status, which can, in turn, lead to more accurate cost predictions during the tendering process, thus minimising the chances of unforeseen expenditure in the future.

While it may be tempting to choose a contractor based on cost alone, this approach can often lead to difficulties during project delivery and result in higher overall costs. My interviews with sub-contractors underscored the importance of considering a range of other factors. It is important to recognise the benefits of working with skilled companies that not only have the necessary expertise but also embrace technology and collaboration as part of their work ethos. Such companies can provide invaluable support during the delivery of the project and save costs during construction through their efficient practices.

Before making a final decision on the choice of subcontractors, it is highly recommended to con-sult your digital construction lead, whose advice can be crucial in assessing a contractor's ability to meet the project requirements. It is also important to maintain open communication with the M&E contractor. Any questions or concerns over the project requirements should be addressed during the tender stage. This proactive approach will help to ensure that all parties have a clear understanding of the project requirements and expectations, thereby contributing to a smoother project execution and successful completion.

My interviews with subcontractors also highlighted the crucial role of an on-site leader, typically a technical services manager (TSM), in the successful implementation of BIM. This individual is responsible for ensuring the timely delivery of information, addressing design clashes and promot-ing digital construction best practices. As part of this role, the TSM proactively identifies potential design issues, mitigating delays and cost overruns, and fosters a culture of collaboration, guiding the team to optimise the use of BIM technologies. This results in a unified approach, maximising the benefits of the design model and leading to improved project outcomes.

The success of projects can be boosted by implementing a few key strategies. One of these is mak-ing it a practice to conduct routine discussions with subcontractors and design consultants. These conversations can provide invaluable insights into your current operational methodologies, making it possible to pinpoint potential weaknesses and implement improvements. Furthermore, I recom-mend providing continuous support to those who express an interest in enhancing their understand-ing of BIM. This support can take various forms, such as complimentary workshops, providing mentorship and advice and creating opportunities for collaboration on BIM-centric projects.

By demystifying the process and tools associated with BIM, and by highlighting its long-term benefits, we can encourage all stakeholders to contribute more effectively to the project's success. Because BIM has proven to be advantageous to all parties involved, facilitating designers and subcontractors in meeting client requirements without encountering difficulties caused by a lack of skills or experience can enhance this success further. Moreover, it would be beneficial to nurture lasting relationships with the right designers and subcontractors, who align with your vision of transformation. This approach can foster a sense of commitment and willingness to adopt BIM in their operations, leading to a higher rate of successful project outcomes.

Table 3.1 aims to outline actions to consider, addressing the common challenges and support requirements identified during interviews with supply chain members. The objective is to mitigate

Table 3.1 Checklist: actions with supply chain

☑ Engage M&E contractors early and, when necessary, other key contractors during the design development phase (RIBA Stages 3 and 4) to enhance the design with their expertise.

☑ Establish a clear understanding of deliverables and responsibilities between design consultants and subcontractors.

☑ Provide a BIM Execution Plan, complete with detailed asset information requirements, in the tender documents, to ensure clarity and understanding of BIM requirements.

☑ Supply additional resources, such as surveys, models and asset data, to ensure that all necessary information is available for review and coordination by the supply chain.

☑ Ensure effective communication with all task teams, especially if certain information is not available, to prevent any misunderstandings or delays.

☑ Clarify project requirements comprehensively at the tender stage by working closely with the team to avoid any assumptions and resolve any concerns.

☑ Allocate ample time for design review during the tender stage, for comprehensive review of project specifications and the coordinated design, to minimise assumptions and extra cost charged by the subcontractors to cover them for any risks.

☑ Establish feedback loops between the lead appointing party and the different task teams to ensure continuous improvement and coordination.

☑ Schedule review periods during the procurement stage for such activities as design coordination, to ensure that all designs are coherent and meet project requirements.

☑ Finalise design coordination before initiating on-site activities, to prevent any potential rework.

☑ Specify and clearly define project stage deliverables, referencing such industry standards as RIBA and BSRIA, to ensure clarity and standardisation.

☑ Assign a dedicated on-site leader, such as a design manager or technical services manager, to ensure that the project stays on track and any conflicts are resolved in a timely manner.

☑ Focus on timely information sharing: ensure that the designated on-site leader prioritises timely information sharing and conflict resolution to keep the project on schedule.

challenges that might arise during the transition from the preconstruction to the construction stage, thereby aiding in the successful implementation of BIM from the outset.

3.4.   Support your supply chain and work with them

Engaging with various task teams at the tender stage requires a tailored approach. Each team has unique requirements and it is imperative that these are acknowledged and addressed within the tender enquiry. These requirements should be clearly delineated and effectively communicated, specifying the information to be provided by each task team as part of the tender process.

Avoid overwhelming task teams with extraneous information, as this could lead to confusion. Such confusion might, in turn, result in inflated fees, as task teams might need to account for unknown risks associated with the lack of clear information. To mitigate this risk, it's advisable to share only the information that is directly pertinent to the project at hand.

Ensuring that task teams fully grasp the project requirements and can meet expectations is fundamental. Providing the necessary BIM documentation, and promptly addressing any concerns or queries related to this documentation, is key. Early engagement with the digital construction lead of the business can facilitate this process.

Lastly, it's noteworthy that companies participating in a digital construction project can be classified into three categories based on their responsibilities. Understanding these categories, as shown next, is essential when selecting the right contractor for your project. By adhering to this guidance, your team can effectively classify task teams based on their project responsibilities and scope. This will enable you to provide them with the appropriate tender information and make appointments accordingly.

Supply chain classification

Group 1
- Develop 3D designs, assist with design coordination and provide structured data, including health and safety (H&S) and operational and maintenance information.
- Collaborate using the common data environment appropriately.

Group 2
- Supply goods required for the project but do not typically provide a coordinated design model.
- Although a 3D design may not be required, it must provide the necessary data to complete the structured asset data, as well as the operational and maintenance documentation.
- Use the common data environment to access and share project information.

Group 3
- Do not offer design services or asset data (e.g., a demolition contractor or plasterer).
- Access and share project information, such as H&S and operational and maintenance details, using the common data environment.

Training initiatives
As you approach the selection stage of your project, it is important to consider the roles of your supply chain partners and their readiness to embrace digital construction methods. With the increasing use of BIM, it is imperative that your subcontractors are equipped with the necessary skills and knowledge to effectively follow the appropriate process and use the technology. Therefore, you have a moral responsibility to offer support and guidance to the different members of your supply chain, if they are not yet on board with digital construction but have performed well on previous non-BIM projects and have the correct culture. They might just need a push in the right direction to help you deliver the best possible outcome on your next project.

It is therefore important to expand your communication strategy to include your subcontractors and ensure that they are well-informed about the benefits of digital construction and your expectations for the use of digital construction in your projects.

Additionally, you should consider providing bespoke training to your subcontractors, tailored to their individual needs and expectations. This will help to bridge any skills gaps they might have and ensure that they are able to utilise digital construction methods effectively in executing your project. With the right support and training, your subcontractors can become valuable assets to your project, contributing to its success and helping you deliver a high-quality outcome.

A capable and knowledgeable supply chain is essential for the successful delivery of digital construction projects, and it is your responsibility to ensure that your supply partners are equipped with the skills and knowledge necessary to contribute to the success of these projects.

To support your supply chain partners, particularly in the case of a large network, you may want seek out a partner with a similar vision and culture that you can trust to provide high-quality training and support. For example, in 2016, I made the decision to partner with BIMBox and held 14 dedicated digital construction workshops for our supply chain partners. These workshops were attended by over 80 different companies within our supply chain and provided a valuable opportunity for attendees to gain a deeper understanding of our company's digital construction culture, process and values at different stages of the design process and final project delivery.

In addition to these workshops, we also instituted BIM assessments for each group as part of the tender enquiry process. These assessments were designed to understand the skills, gaps and culture of each business determining their specific requirements for the projects at hand while identifying areas needing further support. This process was not only instrumental in tailoring the necessary assistance to bridge identified skill gaps but also served as a crucial method for updating our database. This update allowed us to more effectively pinpoint partners who were optimally equipped to deliver BIM projects at the time. Accurately presenting this information to potential clients is crucial in the tender process; doing this helped to ensure that we selected the right supply chain partners for our projects. Please note that although these assessments are very useful and align with the required approach as part of ISO 19650-2 (BSI, 2021), capturing such performances and having the relevant information recorded and available for your teams will help you make an even better selection of the appropriate task team to involve in your project.

The goal of our workshops was to create an environment of learning and collaboration for our supply chain partners, with a focus on individual development and growth. The workshops were designed with a limited number of attendees to facilitate meaningful conversation and open dialogue between participants from various trades and groups. This interactive approach helped them understand their distinct responsibilities in delivering BIM projects and the impact of their roles on day-to-day activities. I wanted to provide opportunities for our supply chain partners to share their concerns and frustrations, as well as to receive guidance and advice from industry experts. Furthermore, those who required more in-depth training and support were offered bespoke in-house training sessions, through BIMBox, to help them transform their businesses and further develop their digital construction skills.

Investing in the development of digital skills among your supply chain partners not only benefits your own business, but it also supports the growth and development of the wider digital construction industry.

Perhaps the aspect that makes me most proud of working in this sector is the opportunity I've had to help others broaden their perspectives, all with the best of intentions to provide support. Through collaborative workshops, we've all learned together and I've been able to guide numerous teams and clients towards adopting a digital construction approach. Although not everyone has embraced our recommendations, many have. It's particularly rewarding to see subcontractors who were initially resistant to change become market leaders with strong BIM capabilities. Witnessing these transformations makes all the effort and challenges truly worthwhile.

3.5. Define and communicate your BIM expectations

In addition to implementing a digital construction policy to foster a digital construction culture within your organisation, it is crucial to establish a clear definition of your BIM approach. While your specific BIM approach will be shaped by the project's client requirements and expectations, it is essential to have a well-defined BIM strategy for projects where the client has not expressed a specific interest in implementing BIM, but your business is committed to following a standardised approach.

The following guidelines can assist you in defining your BIM approach for your business. However, please note that these examples are meant to provide general guidance and should be tailored to your unique business needs.

Documentation

A project-specific BIM Execution Plan, developed in collaboration with the various task teams involved in the project, must be established during the tender stage, irrespective of the client's expectations. This plan is necessary in aiding the supply chain in understanding the detailed BIM approach for each project and ensuring consistency in information production across different task teams throughout the project.

The BIM assessment

As part of the tender stage, the task teams under consideration for the project must undertake a BIM assessment, which entails providing comprehensive information regarding their BIM capability and capacity. They should also address any concerns or enquiries regarding the BIM requirements by consulting the digital team. This information should be included as an integral part of the tender response and carefully evaluated before making the appropriate partner selection for the project.

Information management

It is imperative that all project information adheres strictly to the designated naming protocol specific to the project. Furthermore, such information should be shared through the common data environment, following the project-specific workflows for information approval.

Each task team is responsible for submitting the task information delivery plan (TIDP) based on the project programme before sharing any information using the common data environment.

The team members must strictly refrain from accepting or acknowledging any information sent by email; it is necessary to emphasise the shared responsibility to uphold the importance of and advocate for the use of the common data environment as the exclusive and only acceptable means of sharing information. Similarly, task teams must adhere to this approach using the task information delivery plan (TIDP); no team is to proceed with work without providing the TIDP, which is to be consistently updated throughout the project.

Design process

The design development must be carried out in a 3D environment, utilising BIM authoring tools, with a crucial emphasis on generating all 2D drawings from the latest coordinated federated model.

The practice of converting issued 2D information to a separate 3D model is strongly discouraged and strictly forbidden.

To ensure the integrity of the design, the graphical information and data within the model should be meticulously accurate and well-coordinated. This guarantees that it is suitable for the specific level of information required at each RIBA stage and aligns with the defined scope of work.

Coordination

The lead designer assumes the responsibility of overseeing and managing design coordination throughout the project. While the lead designer takes charge of design coordination using the models, it is crucial to stress that task teams each have individual accountability for ensuring the quality, suitability and spatial coordination of their own content.

It is highly important to address coordination issues as early as possible, and design team meetings provide an opportunity to discuss key issues. To facilitate effective communication and ensure timely problem resolution, the teams involved in the project are accountable for utilising the provided digital tools. These tools enable seamless communication of design issues and help expedite the resolution of identified problems.

Updates

It is essential to accurately incorporate agreed design changes in the model, drawings and all project information at each stage, including site discoveries. Clear responsibilities and fees for these updates need to be defined at the tender stage.

Maintaining consistency and adherence to the accepted design is fundamental during on-site construction, where it is imperative that the latest accepted drawing issued for construction, extracted from the model, is followed. The precision and reliability of the model, drawings and project information collectively play a vital role in supporting the client during the operations and maintenance stage. This comprehensive and accurate information is instrumental in facilitating efficient facility management and providing the necessary data for effective decision making.

Deliverables

The designated task teams responsible for design, as outlined in the BIM Execution Plan, are accountable for delivering:

- the design model, in its native file format;
- the design model, in IFC-SPF;
- PDF and 2D CAD (.dwg) versions of the drawings extracted from the model;
- COBie information exchange at the completion of each RIBA stage.

The frequency of information exchange will be specified in the BIM Execution Plan, considering the project requirements and stage, in agreement with the delivery team. These COBie deliverables should be issued promptly as they become available and should not be consolidated into a single bundle at the end of the stage. It is the responsibility of each task team to input the mentioned data into the common data environment.

Data

If the appointing party has requested COBie, or other structured asset data requirements, in the tender documentation but hasn't specified the asset information requirements, it's crucial to seek clarity. In such cases, it's important to obtain specific guidance on the required data and any particular format requirements during the tender stage.

This clarity not only facilitates verification and validation of the data during project delivery but also ensures that the data are valuable. Moreover, it helps task teams to understand expectations and address any concerns right from the tender stage.

Equally, if the client has not explicitly asked for COBie data, it's vital to consult your company's policy. Consider whether collecting this type of data will align with your business objectives. Remember, that the vast amounts of data collected during a project should have a supporting strategy. Amassing data without a clear plan only complicates processes without adding value.

Throughout the delivery of the project, each task team is accountable for adhering to their appointment and supplying accurate and correct data in the specified format.

Table 3.2 provides a high-level overview of the common assets that are typically included as part of asset data delivery. However, individual projects will have their own asset data requirements, guided by the needs of both the appointing party and the lead appointed party.

Table 3.2 Example maintainable assets list (continued on next page)

Access control	Duct temperature sensor
Actuator	Electric flow storage device
Air diffuser	Emergency exit sign
Air handling unit	Emergency stop button
Air quality sensor	Energy monitoring system (EMS)
Alarm	Evaporative cooler
Attenuators	Evaporator
Boiler	Extractor fan
Break glass point	Fan coil unit (FCU)
Building management system (BMS)	Fire alarm control panel
Busbar	Fire Suppression Terminal
Card reader	Flow meter
Chillers	Flow switches
Closed-circuit television (CCTV) cameras	Gas detection
Compressor	Immersion temperature sensor
Condensers	Isolating valve
Controller	Lighting control panel
Cooling towers	Luminaires (including emergency)
Damper	Passenger and goods lift
Data and telecoms	Presence detector
Distribution boards	Pressurisation unit and expansion vessel
Doors	Pumps

Table 3.2 Continued

Radiator	Transformer
Smoke damper	Thermostat
Smoke detector	Uninterruptible power supply
Socket outlet	Valve
Sprinkler heads	Waste terminal
Tank	Window

The purpose of Table 3.3 is to clarify the common data requirements for each asset, using COBie as a reference format. It's important to note that these data requirements may vary, depending on the specifics of the project. The table covers the data fields required within each tab, identifies who is responsible for providing the data and specifies at what stage the data should be supplied.

Ensure that you clarify with the appointing party the format requirements, nomenclature and length constraints for each data input, such as type names, component names or spaces.

Another crucial aspect to consider is the 'Attribute' tab. Although not frequently requested, this tab holds essential information for the operation and maintenance of assets. Each asset has unique maintenance needs; therefore, having specific data in this tab for the relevant assets is vital. For this reason, input from the facility management team is crucial in determining which specific data entries are required.

To conclude this section, it's important to emphasise that the responsibilities outlined are guidelines and can change based on the project's procurement and appointments. For example, the M&E design consultant will be responsible for providing data at Stage 4. However, if changes arise during on-site execution, the M&E subcontractor might need to update the data during Stages 5 and 6.

Table 3.4 outlines key topics that should be discussed and agreed with task teams during the tender stage. The goal is to clarify project requirements and eliminate any ambiguities. This ensures that the task team delivers the most effective and accurate tender response tailored to the specific project requirements. Doing this helps prevent disputes that may arise when the project starts, which could delay project delivery.

Table 3.3 Data requirements (continued on next pages)

COBie tab	COBie data filed	Stage required	Responsibility
Contact	All fields within this tab	3	Lead designer
Facility	All fields within this tab	3	Lead designer
Floor	All fields within this tab	3	Lead designer
Space	All fields within this tab	3	Lead designer
Zone	All fields within this tab	3	Lead designer

Table 3.3 Continued

COBie tab	COBie data filed	Stage required	Responsibility
Type	Name	4	Design consultants
	Created by	4	Design consultants
	Created on	4	Design consultants
	Category	4	Design consultants
	Description	4	Design consultants
	Asset type	4	Design consultants
	Manufacturer	5	Subcontractors
	Model number	5	Subcontractors
	Warranty guarantor parts	5	Subcontractors
	Warranty duration parts	5	Subcontractors
	Warranty guarantor labour	5	Subcontractors
	Warranty duration labour	5	Subcontractors
	Warranty duration unit	5	Subcontractors
	Expected life	5	Subcontractors
	Duration unit	5	Subcontractors
	Nominal length	4	Design consultants
	Nominal width	4	Design consultants
	Nominal height	4	Design consultants
Component	Name	4	Design consultants
	Created by	4	Design consultants
	Created on	4	Design consultants
	Type name	4	Design consultants
	Space	4	Design consultants
	Description	4	Design consultants
	Installation date	5	Subcontractors
	Warranty start date	5	Subcontractors
System	Name	4	Design consultants
	Created by	4	Design consultants
	Created on	4	Design consultants
	Category	4	Design consultants
	Component names	4	Design consultants
	Description	4	Design consultants
Document	Name	6	Lead appointed party
	Created by	6	Lead appointed party
	Created on	6	Lead appointed party
	Category	6	Lead appointed party

Table 3.3 Continued

COBie tab	COBie data filed	Stage required	Responsibility
Document	Approval by	6	Lead appointed party
	Stage	6	Lead appointed party
	Sheet name	6	Lead appointed party
	Row name	6	Lead appointed party
	Directory	6	Lead appointed party
	File	6	Lead appointed party
	Description	6	Lead appointed party
Attribute	Name	4	Design consultants
	Created by	4	Design consultants
	Created on	4	Design consultants
	Category	4	Design consultants
	Sheet name	4	Design consultants
	Row name	4	Design consultants
	Value	4	Design consultants
	Unit	4	Design consultants
	Description	4	Design consultants
	Allowed values	4	Design consultants

Table 3.4 Checklist: key BIM topics to be discussed and agreed with task teams
(continued on next page)

Project documentation
- ☑ Establish a project-specific BIM Execution Plan during the tender stage.
- ☑ Collaborate with various task teams to develop the BIM Execution Plan and the risk register.
- ☑ Ensure that the plan is clear and aids the supply chain in understanding the BIM approach.

BIM assessment
- ☑ Undertake a BIM assessment for task teams under consideration.
- ☑ Provide comprehensive information on BIM capability and capacity.
- ☑ Address concerns or enquiries about BIM requirements by consulting the digital team.
- ☑ Evaluate responses before partner selection and include BIM assessment summary details in the tender response.

Information management
- ☑ Strictly adhere to the designated naming protocol for the project.
- ☑ Use the common data environment for sharing information.
- ☑ Task teams to submit a task information delivery plan (TIDP) based on the project programme.
- ☑ Clarify that information sent by email will not be accepted or acknowledged.

Table 3.4 Continued

Design process
- ☑ Carry out design development in a 3D environment using BIM authoring tools.
- ☑ Agree on the detailed responsibility matrix.
- ☑ Generate all 2D drawings from the latest coordinated federated model.
- ☑ Ensure that graphical information and data are accurate and well-coordinated.

Coordination
- ☑ Appoint lead designer to oversee and manage design coordination.
- ☑ Address coordination issues as early as possible.
- ☑ Utilise provided digital tools for effective communication and problem resolution.

Updates
- ☑ Accurately incorporate approved design changes at each stage.
- ☑ Ensure that on-site construction adheres to the latest accepted drawings.
- ☑ Support the client during the operations and maintenance stage with accurate data.

Deliverables
- ☑ Deliver the design model in its native file format.
- ☑ Deliver the IFC-SPF.
- ☑ Provide PDF and 2D CAD (.dwg) versions of drawings extracted from the model.
- ☑ Issue COBie information exchange at the completion of each RIBA stage.

Data management
- ☑ Obtain clear guidance on COBie data requirements if not specified by the client at the tender stage.
- ☑ Adhere to company policy for data capture if COBie data is not explicitly requested by the client.
- ☑ Collect and validate COBie data for maintainable assets and assets related to fire safety.

BIBLIOGRAPHY

BSI (2021) BS EN ISO 19650-2:2018 & Revised NA: Organization and digitization of information about buildings and civil engineering works, including building information modelling (BIM). Information management using building information modelling. Part 2: Delivery phase of the assets. BSI, London, UK.

HMG (Her Majesty's Government) (2022) Building Safety Act 2022. The Stationery Office, London, UK.

Amador Caballero
ISBN 978-1-83549-446-2
https://doi.org/10.1680/iceedc.9446204

Chapter 4

Artificial intelligence and data analytics

4.1. Introduction

In the construction industry, it is a requirement for companies to deliver exceptional work within budget and according to strict deadlines while also offering innovative and sustainable solutions. As highlighted in Chapter 1, this sector often operates on tight profit margins, where even minor errors can significantly impact a project's success and the company's overall health. In this high-pressure environment, where large volumes of work are managed within short time frames, AI and data analytics present a significant opportunity to enhance efficiency and work quality. Effective risk and opportunity management is key to success, sustainable growth and retaining a competitive edge. Here, AI and data analytics can prove invaluable.

The adoption of AI and data analytics is increasingly becoming a fundamental strategy for companies to stay competitive. However, caution is essential when implementing these technologies; it's important to consider the maturity of the AI solutions and not to jump in too quickly. While there may be quick wins, having the right strategy from the outset is crucial.

The introduction of AI solutions usually demands significant investment, which may be challenging to justify to a board of directors. Additionally, gathering an appropriate database that is both reliable and insightful, to enable the business to make more informed and strategic decisions, takes time. It's essential to clearly outline the problems and demonstrate how the benefits can solve them, focusing on how the investment can improve productivity and streamline team operations. In my view, AI can enhance performance, optimise processes and increase business turnover without additional overheads. However, the extent of these benefits depends on the precise application and deployment of AI within your organisation.

Thus, integrating AI should be a cornerstone of your strategic and business transformation roadmap. As mentioned in the discussion on FOMO in Section 2.3, while it's useful to monitor competitors, decisions should not be based solely on their actions. What works for them might not work for you. You must ensure that you have the right strategy, prioritising the innovations that should be implemented, and effectively communicating this approach throughout the company.

Success hinges on a thorough analysis of existing methods to identify areas that would most benefit from these technologies. This will reveal new capabilities, leading to improved results across various aspects of construction, from project selection and resource allocation to design methods. Instead of a top-down approach, where only management makes decisions, a bottom-up strategy could be more effective for implementing AI. This involves identifying concrete, quick-win opportunities for AI and data analytics. Start by focusing on areas where immediate and significant operational improvements can be made, making it easier to justify the initial investment to your board of directors.

When considering the use of data analytics, it's important to note that data collection is standard practice in the construction industry. However, merely accumulating data will not yield the desired outcomes. Data must be transformed into actionable insights or knowledge to become useful for the business. Adopting an AI and data analytics mindset from the outset will enable stakeholders to discern patterns, learn from past projects and adjust their strategies accordingly.

By fully understanding and leveraging AI and data analytics, we can elevate industry standards and stimulate innovation. Emphasising a cultural technological approach is key for this transformation. Developments in AI, data analytics and robotics will increasingly shape the industry. Keeping abreast of these changes will enable experts to discover new efficiencies, improve decision making and exceed project expectations. A forward-looking approach to AI and data-centric solutions sets companies on a path to long-term success, differentiating them from competitors.

Incorporating AI in the construction domain should be seen not as a threat to existing job roles but as an opportunity. Data analytics and AI can handle tedious and repetitive tasks, allowing human beings to concentrate on areas where they add unique value.

4.2.   Promote automation

What is automation and why do we need it?

Automation involves using technology to perform tasks that were traditionally done manually, with the aim of reducing human labour, streamlining operations and eliminating repetitive and monotonous tasks. In some cases, automation not only enhances existing procedures but also facilitates the introduction of new ones, thereby improving efficiency, productivity and safety.

As previously highlighted, the construction industry faces significant challenges, including a severe shortage of skilled labour, declining productivity and limited profitability. These issues have adversely impacted many construction businesses, pushing them towards financial strain and, in some instances, leading to liquidation. This troubling trend highlights urgent, deep-rooted problems that require immediate and innovative solutions for sustainable growth in the sector.

In the face of these challenges, automation stands out as a promising solution. It has the potential to transform the construction industry by enhancing efficiency, reducing reliance on manual labour, increasing productivity and improving company profitability.

According to a comprehensive study commissioned by the Midwest Economic Policy Institute (MEPI), *The Potential Economic Consequences of a Highly Automated Construction Industry* (Manzo *et al.*, 2018), nearly half (49%) of all construction tasks could be automated. While the impact might vary between trades, this report indicates that operating engineers, along with painters and cement masons & concrete finishers, could be at higher risk of automation. These findings were based on predictions made by McKinsey & Company in 2017. Considering the rapid technological advances since then, it's likely that an even larger number of roles within the construction industry could be automated in the near future. Table 4.1 is based on findings from the MEPI study. It offers a visual summary that shows how certain trades in the construction sector could be affected by automation. This table serves as a valuable resource for understanding how technology might transform various aspects of the industry.

However, it is crucial to interpret the ascendance of automation not as an alarming trend but as a significant opportunity to augment productivity and efficiency. By taking over repetitive and

Table 4.1 Automation potential by construction trade

Construction occupation	Automation potential: %
Construction trades workers	49
Construction labourers	35
Carpenters	50
Electricians	42
Plumbers, pipefitters and steamfitters	50
Operating engineers	88
Painters	90
Cement masons and concrete finishers	88
Sheet metal workers	39
Roofers	31
Interior designers	12
Surveyors	56
Cost estimators	14
All other construction occupations	35

Source: Manzo *et al.* (2018) and Johnson (2017).

tedious tasks, automation enables the human workforce to redirect their attention to more intricate and high-value tasks. This shift can engender a more engaged and motivated workforce, opening up avenues for swift growth and innovation in the sector.

Automation, by eliminating monotony, risks and optimising productivity, presents an exciting opportunity to rejuvenate the construction industry and ensure its continuous progression. This transition would not only improve industry dynamics but could also pave the way for a more sustainable and resilient future for the construction sector.

Beyond just robots

When we think about automation, for many, the first image that often comes to mind is robots. However, in the context of the construction industry, the applications and potentials of automation are much broader than this. Indeed, the term 'automation' doesn't just refer to robots; it encompasses a wide range of advanced technologies. These include, but are not limited to, artificial intelligence (AI), drones for site surveying and monitoring, 3D printing for crafting detailed building elements, as well as immersive augmented reality (AR) for improved design visualisation and collaborative design. These technological advances are capable of analysing large amounts of data intelligently and supporting us to make faster and well-informed decisions.

You may already be aware of some practical examples of how automation is advancing the business side of construction. One revolutionary development is autonomous or self-driving vehicles.

These vehicles, such as bulldozers, excavators and dump trucks, can operate independently on construction sites, providing increased efficiency and productivity.

Several pioneering companies, including Built Robotics, Caterpillar, Doosan and Volvo, are at the forefront of developing and testing autonomous heavy machinery for construction sites. While we are still in the early stages of implementing fully automated machinery, several semi-automated solutions are under development. However, it's important to note that fully autonomous construction machinery is not yet widespread. These vehicles come equipped with various features, including an all-weather enclosure, proximity radar, 360° cameras, Global Positioning System (GPS) technology and a powerful liquid-cooled computer system.

Automation also plays a significant role in the manufacture of prefabricated building components and modules. Automated systems are used in factories to produce these components, guaranteeing greater precision and quality control. Once these components are ready, they are transported to the construction site and assembled quickly. As highlighted during my interviews with supply chain teams (described in Chapter 3), this process significantly reduces on-site construction time, thereby increasing overall project efficiency.

Beyond manufacture, automation is employed in construction site monitoring and management systems. This includes the utilisation of Internet of Things (IoT) sensors that monitor construction progress, track site conditions, track assets and resource usage, and ensure safety compliance. Predictive maintenance is another important application of automation. Driven by data analytics, it can anticipate and prevent system disruptions, substantially reducing down time and enabling quicker design development.

Moreover, automation is employed to operate heavy machinery, such as cranes, increasing both safety and profitability. An example of this is the use of UltraWis cranes on site, which are operated using automated control systems.

These examples merely scratch the surface of the potential of AI and automation in the construction industry. As technology continues to advance, it is expected that we will see even more transformative uses of automation in this field.

Robotic innovations

If we closely examine various instances of how robotics are being integrated in the construction industry, it becomes evident that this sector is evolving at a remarkable speed. A primary advantage of employing robots is their ability to enhance efficiency and address the prevalent issue of skilled labour shortages. More importantly, robots can undertake tasks that could potentially pose risks to human beings. For instance, they can operate in dangerous environments or handle hazardous substances that are unsuitable for human interaction. Broadly speaking, the deployment of robots serves to bolster the workforce while simultaneously elevating productivity. I'd like to use this moment to discuss some intriguing applications of robotics that have piqued my interest.

Bricklaying robots

Robotics technology has revolutionised bricklaying with the introduction of robots like the Hadrian X by Fastbrick Robotics and the Semi-Automatic Mason (SAM) by Construction Robotics. Capable of laying up to 1000 bricks per hour, these robots significantly alleviate the physical strain on human workers and improve the speed and quality of construction.

Moreover, tools like the Material Unit Lift Enhancer (MULE) by Construction Robotics are designed to handle and place heavy materials on construction sites, substantially reducing the risk of worker injuries.

Humanoid robots and semi-autonomous systems

Humanoid robots, like the HRP-5P, crafted by Japan's National Institute of Advanced Industrial Science and Technology (AIST), have been engineered to assist with drywall installation tasks. Despite being in ongoing development, these humanoid robots hold the promise to transform a number of sectors. Meanwhile, semi-autonomous systems, such as the Canvas robot, have been developed to perform such tasks as applying finishing compounds and sanding drywall. Similarly, Transforma Robotics' Pictobot can paint at a speed that is four times faster than traditional methods.

These technological advances not only reduce the risk of repetitive strain injuries but they also significantly improve overall work efficiency, leading to projects being executed faster and more efficiently.

Robots for routine inspections and site surveys

Without doubt, such robots as Boston Dynamics' Spot, along with various drone technologies, have become extremely popular. It has become the norm at digital construction-related events to have them in attendance. These technologies are utilised for regular inspections, site surveys, safety monitoring and hazard identification. While they have numerous potential uses, these robotic tools are primarily designed to gather consistent and accurate data. This not only enhances productivity but it also significantly reduces the probability of accidents happening on construction sites.

Other robotic innovations

The industry has also seen the introduction of numerous other innovative robotic solutions. For instance, Dusty's FieldPrinter can autonomously print full-scale models onto construction surfaces in a fraction of the time it takes a manual layout crew. Similarly, Husqvarna Demolition Robots can increase productivity and safety by over 20%.

A world of possibilities has been opened up by 3D printing technology in the construction industry. An excellent example is the stainless steel bridge printed by MX3D in Amsterdam. On-site 3D printing allows for the creation of bespoke designs that might be challenging to achieve using traditional manufacturing methods. It also helps reduce waste and minimise transportation needs.

Welcoming the future of construction

The construction industry can benefit greatly from adopting technology and automating certain tasks. By implementing these methods, projects can be completed faster and be more profitable, with a greater emphasis on safety, resulting in better outcomes and ultimately greater success. This transition to automation, however, calls for a willingness to embrace change and to maintain an open mind. With these attributes, the rewards can be substantial, resulting in a win-win situation for everyone involved in the industry. The future of construction is indeed on the horizon, and it's looking brighter than ever.

Despite these exciting prospects, many people fear that AI could threaten their job security, as previously mentioned. The real issue isn't that AI will take their jobs, but rather that someone more adept at utilising artificial intelligence could outcompete them. However, as technology advances, exciting possibilities are on the horizon that offer an opportunity to fundamentally rethink the nature of work itself. The advent of AI should not be viewed merely as a challenge or a threat to existing jobs. Instead, it's a chance to innovate the way we perform our roles. The impact will be profound, leading to a shift in how we engage with our jobs on a day-to-day basis.

4.3. Use of AI to accelerate design decisions

The integration of AI into business strategies is currently a hot topic that's generating a lot of discussion. When people raise concerns about the threat of AI to employment in various sectors, I'm reminded of an experience from years ago. We introduced a tool that could generate quantity take-offs directly from 3D models. Initially, there was resistance from the estimating team, fuelled by fears that the technology would replace human input. However, the outcome was quite the opposite. The tool saved the team time and allowed them to add additional value to the estimating processes.

I understand the concerns about automation and job displacement, but my experience has shown that technology often empowers us to perform better in our roles rather than threatening them. So, let's shift our perspective. Instead of viewing AI as a potential threat, we should see it as a powerful tool designed to augment our capabilities. By embracing this technology, we set the stage for a more productive, creative and innovative future for both individuals and the industry as a whole.

It's abundantly clear that integrating AI into operational workflows has broad benefits. This is particularly evident in the construction sector and the design process, which have traditionally been bogged down by tedious tasks.

Traditionally, architects invest vast amounts of time creating requisite drawings and documentation. This often limits their capacity to produce innovative and effective design solutions that meet client needs and ensure functional efficiency.

The architectural and engineering industry has witnessed significant evolutions over the years. Transitioning from traditional drawing boards to computer-aided design (CAD) in the 1980s and then to 3D designs marked substantial progress, enhancing coordination and communication throughout the design phase. Now, AI is introducing a new era of innovation by optimising design processes. The shift to AI is happening even faster, highlighting the urgency for professionals to swiftly adopt new innovations to stay competitive.

While current tools in the Architecture, Engineering and Construction (AEC) industry often fall short of meeting designers' needs and face challenges in data exchange, new AI tools are revolutionising the market.

The adoption of AI and machine learning technologies, a subset of AI, is on the rise and having a significant impact across different phases of the design process. Generative AI serves various purposes, such as creating new content and aiding in decision making. It learns from existing data and can be trained with new information. For instance, it can streamline building design by automating repetitive tasks. Additionally, it can check designs against building regulations or be programmed to verify other specific requirements, thereby facilitating compliance and approval processes.

Additionally, machine learning algorithms can generate various design alternatives based on set parameters. This flexibility allows designers to rapidly explore multiple solutions and choose the one that best meets both the project's objectives and client expectations.

Another area where these technologies excel is in speeding up and improving design coordination tasks. Given their adaptability, these technologies are still in their infancy but have enormous potential to support teams in optimising time and delivering higher-quality outcomes.

In the realm of intelligent design tools, a significant advance has been in the area of structural optimisation using AI. It involves applying artificial intelligence to create lightweight, efficient designs that meet all performance and safety criteria. One example of such a tool is Altair, which aids engineers and architects in arriving at optimal solutions for complex structural problems.

Similarly, SWAPP is an AI-powered, cloud-based application that revolutionises the creation of comprehensive 3D design models and associated documentation, considerably reducing the design phase. This is complemented by tools like TestFit, Autodesk Forma and Hypar, which collectively empower design teams to swiftly plan and systemise building projects. Emerging technologies like text-to-BIM are also gaining momentum and herald exciting prospects for future innovations.

Speckle offers additional benefits by providing unique features for designers, such as data management and automation features, which enhance team communication and efficiency. For those looking to create visually stunning representations, tools like Midjourney or Adobe Firefly come into play, transforming text descriptions into high-quality visuals that could easily be mistaken for photographs. This accelerates the postproduction of images, allowing teams to invest their time more efficiently.

As we can see, by using AI tools, architects have the ability to automate a significant portion of the drafting and documentation process, as well as generate multiple design options. These tools allow them to optimise designs and seamlessly handle tasks such as rendering and analysis. This not only streamlines workflows but also enables architects to focus on the core aspects of the design process, resulting in superior designs and enhanced client satisfaction.

Embracing these AI tools enables businesses to diversify their service offerings, secure more contracts and even create job roles that were previously inconceivable. Moreover, AI applications can significantly elevate operational efficiency and competitiveness, positioning firms favourably in the marketplace.

Interestingly, the utility of AI is not restricted to industry professionals. Platforms like Interior AI enable users to upload room photographs for AI-driven design visualisations. This helps to streamline the design selection process and enhance interactions with clients. Despite its advanced capabilities, AI does not replace professional expertise but rather augments the design process.

As AI evolves and becomes increasingly accessible, its role in design and construction will only grow. While AI will never negate the need for human expertise, it will undeniably redefine how professionals operate, making processes more efficient and creating novel opportunities. The tools are merely indicative of the current market offerings. Given the rapid technological advances, these tools might soon be superseded by even more sophisticated alternatives. The dynamic nature of AI challenges our preconceived notions of traditional practices, heralding an exciting era of innovation in the architectural and engineering sector.

4.4.   A data-driven approach to project selection

The role of data analysis in decision making

In the world of business, and particularly in sectors as fiercely competitive as construction, the significance of employing the right tools and strategies to stay ahead of the curve is a necessity. Among the excess of strategies businesses could adopt, one stands out for its practicality and effectiveness – data analysis. Data analysis has become more than a modern buzzword. It has evolved into a crucial differentiator, aiding businesses in identifying trends, making informed decisions and selecting projects that best suit their capacities and resources.

Decision making in the construction industry, especially when deciding which projects to undertake, is as critical as ensuring their successful execution. A project cannot be taken lightly, as it involves significant financial and resource implications. Therefore, businesses must conduct exhaustive and meticulous analysis of a project's details before committing. A flawed understanding or misjudgement of a project could lead to uninformed decision making, which in turn could result in substantial financial loss and severe damage to the company's reputation.

The tender decision making process in the construction industry is complex and multifaceted. It involves several factors, each of which can significantly influence the success or failure of a project, as well as its economic profitability. Therefore, each of these factors must be evaluated with utmost care and precision before deciding to tender for a project. This evaluation should encompass a wide range of aspects, such as the size and complexity of the project, the sector in which it falls, the location, trusted supply chain partners in the area, the project timeline and the budget. Furthermore, thoroughly understanding the client's needs and expectations is an undeniable necessity. This involves recognising their explicit and implicit needs, understanding their decision making process, and being aware of how they interact with businesses.

Tendering is cost-intensive, highlighting the need for comprehensive analysis before participation. In today's data-centric environment, businesses can harness data from past projects to discern which ones have been profitably successful. Such analyses allow them to channel their resources towards similar projects, in line with their capacities, which promise profitable outcomes.

By employing data analysis, companies can determine which projects, clients, contract styles and other factors align with their competencies. A data-driven approach not only increases the likelihood of project success but also assists in project and client selection. Accepting projects indiscriminately is ill-advised. Proper due diligence must precede any commitment to a project tender. This cautious approach enhances the probability of project success, fostering client satisfaction, repeat business and consistent company growth.

Understanding the role of data in predicting the likelihood of winning a project and identifying the factors that contribute to successful project delivery is an integral part of modern construction business practices. By analysing past performance, businesses can identify their areas of strength and those that need improvement. This evaluation can guide the changes and modifications necessary to maximise success in future projects.

Building trust and maintaining strong relationships with clients and suppliers is another crucial aspect of the process. A positive reputation in the industry, good performance in previous projects, and strong client relationships are critical factors that contribute to future project selection decisions. As such, businesses must ensure that their performance is consistently captured, documented and analysed, as this information can greatly inform future project selection decisions.

The significance of data in the tendering process

The process of tendering a project involves numerous costs, including but not limited to, the allocation of personnel, procurement of materials and various administrative expenses. These costs can vary greatly, depending on the scale of the project and the number of resources required to complete it. Losing several tenders in a short period can be more than just financially damaging – it can also be incredibly frustrating and demoralising, affecting team morale and overall performance. Hence, maintaining a favourable win-to-loss ratio is essential.

To achieve this, organisations must adopt a strategic approach to the tendering process. This involves systematically gathering and analysing data to gain insights into past project experiences. When leveraged correctly, this pool of information can serve as a powerful tool to enhance the chances of success in future tender bids.

Various factors can influence a tender's success. These include project value, location, duration, industry sector and the type of project. Understanding the impact of these variables on the tendering process can inform decisions about whether to tender for a particular project and where to concentrate efforts for the best chance of success.

This learning process involves tracking feedback and the reasons for each won or lost tender. This practice helps identify trends and areas for improvement for future tenders. It serves as a valuable learning opportunity, where organisations can understand what went wrong in unsuccessful bids and how to make necessary improvements for future tenders. For instance, consistently losing tenders because of pricing could indicate a need to revisit the pricing strategy, negotiate better deals with suppliers or review the allocation of internal resources required for project delivery.

Data analytics can bolster business turnover predictions, which are pivotal for informed decision making. Additionally, in such areas as team selection, data analytics prove invaluable. By evaluating the skills, expertise and past performance of team members, organisations can ensure that each project is assigned to individuals best suited to execute it effectively. This can significantly increase the chances of project success, while reducing the risk of costly mistakes or delays.

In our competitive world, data analysis has transformed into an invaluable asset for companies in the construction industry. By harnessing the power of data to predict the likelihood of winning a project, allocating resources more effectively and improving the quality of tenders, companies can significantly enhance their competitiveness and sustainability. With the right strategies and tools at their disposal, businesses can greatly increase their chances of success and achieve their business objectives.

Despite the promising advantages of data analysis, it's essential to remember that this approach's effectiveness largely relies on the quality of data. To analyse trends and make informed decisions requires a large, accurate and comprehensive database. The more variables considered, the more data points are required to accurately analyse trends and make decisions based on past projects.

This underlines the necessity of investing time and resources into building and maintaining a comprehensive database for effective data analysis. Companies should not shy away from dedicating resources to establish a robust data strategy, as the benefits of doing so will accrue in the medium to long term.

4.5. Improve planning and resource allocation

Leveraging historical data for enhanced project planning

Having an accurate and efficient programme is crucial for securing construction bids and setting your business apart from competitors. Typically, the conclusion of the programme proposal occurs near the tender submission deadline, with its complexity leaving little room for further modifications or additional activities, such as logistics and construction visualisations. This tight time frame underscores the necessity for a method to produce programmes faster, to permit further review and validation by the delivery team. By affording more time to additional tasks dependent on the proposed programme, bid success and overall project execution can be significantly enhanced.

Additionally, the accuracy of the programme is vital for the success of a project, as it lays the foundation for the entire construction process. In construction projects, it's common for planners and the rest of the delivery team to make decisions based on expert judgement. However, this can introduce a level of uncertainty, especially if the expertise was not gained within the current company. Overlooked variables, such as internal processes, lack of team expertise or unfamiliarity with the project type, may lead to a number of assumptions during the execution of the programme, resulting in project delays and cost overruns. This can have a detrimental impact on both the project and the company's reputation.

Leveraging data at the tender stage significantly enhances programme accuracy by removing subjectivity based on experience and the analysis of historical project data within the business. This analysis not only refines the predicted accuracy and time estimation for each activity but also quickly identifies delays and their causes from previous projects, enabling the team to account for these in future project programmes. This proactive approach mitigates the risk of project delays and cost overruns, which can notably impact project success and the company's bottom line. Moreover, access to a comprehensive and trustworthy dataset from previous projects allows for better-informed decisions and more effective resource allocation by the planning and operations teams. This targeted approach leads to cost savings and higher project success rates.

Predicting overall programme duration is more challenging than predicting individual activities' durations, owing to numerous variables that can impact the programme. However, a well-established and reliable historical database enables more informed predictions by analysing trends and patterns based on such factors as project type, sector, location and budget. This data-driven approach can lead to more accurate scheduling and better resource planning, ultimately benefiting the company and the project's stakeholders.

While data are a valuable resource, specific project details should still be considered when making decisions, especially for projects that deviate from previous experiences or involve innovative approaches. Knowledge and expertise gained from past projects should be recorded accurately in an easily accessible system, allowing employees to make informed decisions and build on the company's collective experience. This emphasis on data retention and knowledge management is vital for fostering a culture of continuous improvement within the organisation.

Accurate data collection can enhance construction schedules by providing insights into the optimal task sequence for a specific site and uncovering construction solutions that streamline task sequences and improve overall schedules. Identifying potential bottlenecks and inefficiencies early in the planning stage can save both time and money, contributing to a more successful project outcome.

Artificial intelligence and strategic planning for project efficiency

A shift in culture is essential, not only to welcome the application of data and AI tools but also to appreciate the long-term value and benefits that come from comprehensive work planning and adequate time lengths during the preconstruction phase. The preconstruction phase tends to be shorter when dealing with tight programmes. However, giving more attention to meticulous planning, conducting comprehensive surveys, and ensuring detailed design coordination, all while carefully addressing assumptions, significantly enhances on-site efficiency. This integrated approach helps prevent costly errors, unnecessary ad hoc decisions, avoidable reworks, and frustrating delays. Conversely, a lack of adequate time for these interconnected activities can have a detrimental effect on the project's success, negatively impacting its quality and the ability to adhere to budget and schedule constraints.

An existing cultural propensity towards initiating on-site work as quickly as possible can often lead to solving problems on the spot rather than investing in a longer preconstruction phase to mitigate potential issues in advance. Our objective should be to persuade and communicate the benefits of the correct preconstruction programme that will lead to the successful completion of the project to the client, rather than rushing to its initiation.

Thus, it's apparent that, while aggressively fast-paced plans presented during the tendering process may initially appeal to clients, these might not have been rigorously and strategically scrutinised. This lack of scrutiny can impact the execution of the works and the eventual project handover date.

Further, data analysis and AI-driven planning tools can play a significant role in refining the planning process. Several AI tools, like ALICE and Frontline, have the potential to revolutionise planning in the construction industry by offering a host of benefits. These tools efficiently explore different construction alternatives, while taking into account project constraints. This rapid exploration speeds up the planning process and allows planners to focus on selecting and analysing the most optimal path for the project. Having an appropriate plan not only improves the chances of winning a bid but also ensures a smooth project execution, increasing the likelihood of profitability. These technologies promote data-driven decision making, reduce project risks and automate the generation and updating of optimal schedules throughout the project.

These are just early examples of how nascent technology can significantly impact the planning process throughout a project's lifecycle. These insights provide guidance on how you might consider integrating such technology into your business operations while keeping an eye on the evolving landscape of AI solutions geared towards planning support.

Effective resource allocation

Effective resource allocation is a critical aspect of project management in the construction industry. A reactive approach based solely on secured projects can lead to challenges from under-resourced projects, delays and inefficiencies. The ideal strategy involves proactive resource planning, taking into account various factors, such as team skills, seasonality and organisational needs.

Understanding your pool of available resources is crucial for successful project execution. This involves maintaining a list of experienced professionals and building strong relationships with subcontractors. Inadequate planning can result in hasty and suboptimal resource allocation, leading to compromised project quality, low team morale and reduced profitability.

Data analytics and AI are increasingly important in enhancing resource planning and allocation. These technologies sift through large datasets to identify underlying patterns and trends, streamlining decision making processes. A data-driven approach positions businesses to optimise their resource allocation strategies more effectively.

A comprehensive resource planning approach involves several key strategies.

- *Transparent resource planning.* Implementing a visual tool accessible to all team members helps track past demand patterns, current project requirements and the future pipeline. This tool should be updated in real time and informed by data analytics.
- *Resource availability.* Good planning also considers human resource constraints, such as leaves of absence, training commitments and commutes, ensuring realistic and practical resource allocation.
- *Communication and collaboration.* Open communication channels are essential. Tools providing real-time progress updates on site, such as the Buildots platform, help teams make informed decisions about resource allocation.
- *Flexibility.* Businesses must be prepared to adjust resource plans based on changing project requirements or unexpected events. Having a cohesive team and backup options, for instance, trusted freelancers, can be helpful.
- *Expertise and balance.* Skills and expertise must be factored into the allocation process, considering the project's nature, duration and location for not only successful completion but also employee work–life balance.
- *Data-driven decision making.* Incorporating AI and analytics helps businesses make precise, informed choices. Analysing patterns and seasonality enables more accurate predictions to be made about resource requirements, optimising allocation and minimising waste.
- *Continuous improvement.* A commitment to regularly evaluate and refine resource allocation strategies is essential for long-term success.
- *Resource management software.* Comprehensive resource management software offers real-time insights into resource availability and utilisation, aiding in effective project staffing and allocation.

4.6.   Utilising AI for efficient project delivery

Capture the works completed on site

The construction industry requires a great deal of planning and attention to detail to ensure the successful delivery of projects. One aspect that is often underestimated is the importance of having a clear and accurate understanding of the progress being made on site.

Having a clear picture of the project status is essential for making informed decisions about the direction of the project and the actions required to ensure it stays on track. However, there is often a perception gap between what the delivery teams believe is happening and the actual status of the work. This discrepancy can lead to poor progress reports that do not accurately reflect the state of the project. These reports can have a profound impact on the delivery of the project and can even influence the actual completion date. If the perception gap between what is believed to be happening and the reality is not aligned, this can create confusion and make it difficult to determine the next steps.

Moreover, to control and reduce the cost of error in the project, consistent and accurate tracking is vital. Additionally, it not only ensures better management of resources but also helps in compliance with the Building Safety Act 2022 (HMG, 2022).

Producing high-quality progress reports is demanding and requires a great deal of effort and attention to detail. It is not always easy to keep track of the progress of different trades while also managing other daily activities. The complexity of the task is compounded by the fact that the delivery team on site must have a clear understanding of the status of the work and be able to communicate it accurately and effectively.

Therefore, it is vital that the delivery team on site is equipped with the necessary tools and skills to produce accurate, quick and comprehensive progress reports.

In my personal experience, there are two key factors that significantly affect the disconnection between the reality of the progress of works on site and the perception of the delivery team.

- *Interpersonal skills.* The ability to effectively communicate and understand the current state of the project is essential for producing accurate progress reports. In such cases, the team might not have a clear understanding of the position of different trades in relation to the programme, owing to their lack of skills or knowledge, and this can result in inaccurate progress reports.
- *Subjectivity.* Another significant factor that contributes to the disconnect between reality and perception is the subjectivity of individuals. For instance, if different building managers were to work on the same floor or area of a project, it could be possible that they would have differing interpretations of the progress made. The subjectivity in the reports of different individuals based on their expertise and knowledge makes it challenging to accurately determine the real status of the work.

In addition to these two factors, there is often a tendency to be overly optimistic when producing progress reports. This lack of honesty can stem from a fear of criticism for unfavourable progress or a desire to present the project in a positive light.

This tendency to be overly optimistic in these reports can lead to a skewed perception of the actual progress of the project and create confusion and misunderstandings. Therefore, it is imperative that the construction industry embraces technology to gain a more accurate understanding of the progress of the works on site, which will help make well-informed decisions.

The traditional methods of tracking progress and providing evidence of work completed have previously been hindered by the subjective opinions of the various teams involved, leading to a perception gap between what the teams believe is happening and the actual reality of the situation. However, the advent of digital construction has provided a solution to this challenge. By utilising site capture imagery, such as 360° cameras or handheld scanners, it is possible to record and capture the work being done on the construction site as one moves around. This captures a comprehensive view of the progress and allows for real-time analysis of the work completed.

The adoption of site capture imagery can help to eliminate subjectivity and provide a clear and objective picture of the work completed at any given time. This technology enables you to capture images of the site and automatically map them to the floor plan drawings, providing a virtual view of the site at each stage of the project. This is like having a Google Street View of your project, providing numerous benefits to all stakeholders involved.

One of the main benefits of this technology is the time-saving aspect. With traditional site photos, it can be extremely time-consuming to identify the correct image when it is needed, as each photograph must be named correctly, identified on the floor plan and saved in a specific date folder. With 360° cameras and AI, this process is automated, saving time and increasing efficiency.

Furthermore, this technology provides a level of transparency and accountability that is crucial in avoiding disputes and protecting your business in the long run. With the ability to track the progress of the works on site in real time and provide objective evidence, it becomes easier to resolve any disputes that may arise. This makes it imperative that this kind of technology should be mandatory for every project that your company undertakes, regardless of size or complexity.

Drawing on my extensive field experience, I can confidently state that this technology has the power to revolutionise construction project management and execution. Its potential for transformation is undeniable, and it can significantly impact the industry in numerous ways. One notable advantage is its ability to play a vital role in verifying the quality of work performed by subcontractors, ensuring compliance with regulations and minimising conflicts. These points represent a summary of the benefits that can be achieved.

- *Objectivity and accuracy.* This technology eliminates subjectivity from progress reports, providing an accurate and consistent source of information for all project stakeholders.
- *Dispute resolution.* The technology helps resolve disputes related to commercial, quality or safety issues, saving time and money in the long run.
- *Progress comparison.* It allows easy comparison of progress at different stages, enabling project participants to identify areas for improvement.
- *Remote monitoring.* The technology enables progress monitoring from any location, supporting agile work practices and reducing travel needs for enhanced efficiency and sustainability.
- *Increased transparency and confidence.* By offering transparency to clients, it eliminates the need for on-site visits, enabling quick responses to enquiries.
- *3D modelling comparison.* The technology compares 360° images with 3D models, ensuring adherence to the original design and enabling prompt identification of any deviations.

Identify changes and eliminate disputes on site with AI

One technological advance that has captured my attention in the realm of digital construction is the integration of site capture imagery with AI, exemplified by tools like Buildots. Its ability to provide accurate and objective information about on-site progress to delivery teams is what sets it apart.

This innovative technology uses computer vision and machine learning algorithms to analyse data on the construction site in real time based on the 360° images captured during the site walks. It provides an objective, accurate representation of what has been built and what still needs to be done based on the agreed design and construction programme.

As part of the digital construction process, in our efforts to improve project delivery and minimise issues on site, we invest a significant amount of time and energy into coordinating the design using models and conducting coordination reviews. However, all this effort can go to waste if the construction teams don't follow the accepted design.

It's common for subcontractors to make changes on site while installing the works, either to improve or simplify the installation or because of unexpected discoveries in existing buildings. However, these changes from the accepted design can lead to bigger problems in the long run. They can disrupt coordination with other trades, cause delays, increase costs, lower the quality, violate specifications and building regulations, and, ultimately, disappoint the client who accepted the design.

The construction industry frequently faces the problem of subcontractors deviating from the accepted design without proper change management. This issue has a notable effect on both the project timeline and cost. Failing to identify these deviations promptly can have negative consequences beyond the project delivery. It exposes the contractor to potential non-compliance with regulations, leads to lower quality information for clients, resulting in client dissatisfaction, and damages the reputation of all involved stakeholders. Therefore, it's crucial to ensure that subcontractors don't deviate from the design without obtaining approval for new construction revision drawings. Some may prefer to take action on site without waiting for paperwork to be completed, but the consequences of these improvised deviations can be costly, as we've seen before.

There's a need for a cultural change where deviations from the accepted information on site are not only discouraged but penalised. Any deviations from the intended design must be immediately identified and promptly resolved. This ensures that the construction project stays on track and the final product meets the quality and compliance standards set out in the accepted design, specifications and regulations.

Owing to the presence of different trades and the fast pace of work on site, it can be challenging for the site team to notice contractors' modifications to the design. This often results in overlooked errors that are only discovered later when more significant issues arise or when other stakeholders bring them to light. To mitigate these potential problems, tools like Buildots, which utilise AI technology, can be used. This tool diligently monitors every aspect of the building process and sends notifications whenever something is not installed correctly or deviates from the accepted design. This is revolutionising the construction industry by providing real-time updates and significantly reducing the risk of costly mistakes.

Additional benefits of AI

The incorporation of AI in the construction process holds immense potential for revolutionising the industry, offering numerous advantages to both construction companies and their clients. Beyond the already mentioned benefits, for instance, using 360° images to monitor site progress and easily detect changes and errors, the integration of AI takes the technology to new heights. By linking these images with the design model, commercial information and construction programme, the project delivery team gains access to valuable data that empowers them to make informed decisions and swiftly resolve issues.

The implementation of this technology can also help to improve communication and collaboration between stakeholders, including the contractor, subcontractors and project delivery team. By having access to real-time information, all stakeholders can make informed decisions and take proactive measures to resolve any issues that may arise. This approach provides a clear and concise picture of the construction project, allowing all stakeholders to stay informed and up to date on progress and potential issues, leading to fewer disputes with different stakeholders.

Historically, the construction industry has been consistently marred by issues of transparency and the failure to provide high-quality construction record information, leaving clients often dissatisfied and dubious about the accuracy of the information they receive. However, a solution to these persistent challenges has now emerged. Leveraging technology allows for swift identification and on-site correction of any discrepancies from the accepted design. This not only reassures clients about the high quality and accuracy of the information being handed over to them but also enhances trust between the construction company and its clients, a fundamental element for the success of any business.

Objective data allow for more informed decisions to be made and provide a clear understanding of areas that require improvement. Additionally, AI tools use real-time data to predict the completion time of tasks based on previous performance, as well as to determine the resources required to meet project goals. Having a better delivery and identifying errors on time helps you to minimise the needs for revisits, saving time and cost.

Another significant benefit of AI in construction is the ability to reduce the number of snags at the end of a project. Snags can cause significant delays and additional costs, but with the use of AI, any activity not completed correctly during construction will be identified and addressed in a timely manner, avoiding snags at the end.

In addition, AI enables you to make data-driven decisions, reducing the risk of delays and cost overruns, by having the correct amount of resources in the appropriate areas of the project site. With this technology, you can ensure that the construction project is moving forward smoothly and efficiently and that the product meets the highest standards of quality.

Moreover, objective data can also be incredibly helpful in comparing the performance of different trades and analysing the impact of delays for other subcontractors. This information can then be used to efficiently reorganise works and avoid potential time loss. Furthermore, AI technology can be used to quantify the percentage of work completed by each subcontractor and assist in the valuation of the project. This can minimise conflicts and provide visual evidence of work completed, which can be especially valuable in the event of disputes. Additionally, Buildots and similar tools enhance work sequencing and enable workspace to be maximised, leading to faster project completion. Easy access to necessary resources and labour in available areas promotes efficiency and cost-effectiveness. Monitoring unused space is crucial for measuring project productivity and optimising resource and labour utilisation.

The incorporation of AI in the construction field is still relatively new, yet its potential to enhance efficiency, productivity and quality holds much promise. This emerging technology opens up a myriad of possibilities for the future, and it is fascinating to consider how AI could further refine the construction process. I'm of the view that this technology will not only boost productivity and streamline efficiency, thereby providing superior projects to clients, but it could also lessen the necessity for intensive on-site management. This reduction could in turn reduce overhead costs, making projects more cost-effective.

BIBLIOGRAPHY

HMG (Her Majesty's Government) (2022) Building Safety Act 2022. The Stationery Office, London, UK.

Manzo J, Manzo F and Bruno R (2018) *The Potential Economic Consequences of a Highly Automated Construction Industry*. Midwest Economic Policy Institute. https://midwestepi. files.wordpress.com/2018/01/the-economic-consequences-of-a-highly-automated-construc- tion-industry-final.pdf (accessed 21/11/2023).

Johnson, D (2017) *Find Out If a Robot Will Take Your Job*. time.com, 21 April. https://time. com/4742543/robots-jobs-machines-work/ (accessed 17/11/2023).

Amador Caballero
ISBN 978-1-83549-446-2
https://doi.org/10.1680/iceedc.9446205

Chapter 5
Understanding BIM and its foundations

5.1.   Introduction

Building information modelling (BIM) provides clients and construction professionals with better information management and digital tools facilitating enhanced project outcomes and stakeholder collaboration. This chapter delves into the core of BIM, exploring its principles, processes, benefits and the challenges faced in its adoption within the industry.

The drive for change should come more from clients than from regulatory requirements. This chapter covers the BIM Mandates and their impact in the UK, emphasising how crucial it is to communicate the benefits of successful BIM implementation. When clients recognise both the tangible and intangible benefits that BIM offers, the pace of its adoption quickens, even without the need for mandates.

This section explores the high-level process of BIM and sheds light on the most important considerations at each stage. In this context, comparisons are drawn between traditional construction methodologies and the BIM approach. The goal is to demystify BIM and alleviate any apprehensions surrounding its adoption.

Implementing BIM brings numerous benefits. However, the path to BIM adoption also presents various challenges. These range from technical constraints to resistance to change and reflect the complexities inherent in transitioning to digital construction. Therefore, to support this transition and help people embrace change, it's important to keep things simple and ensure that communication is practical and straightforward.

A key section of this chapter focuses intensely on the very core of BIM: robust information management. A thorough review of BIM principles, the common data environment, information approval workflows and naming protocols is essential for maintaining the accuracy and reliability of information throughout the project, ultimately driving its success.

I analyse insights from the PwC (2018) report, which examines the impacts of BIM and measures its benefits. I pay special attention to those that manifest after the handover stage. This helps clients understand the importance of BIM for them and, hopefully, encourages the industry to adopt it.

Lastly, the spirit of collaboration over judgement is fundamental to the success of BIM implementation. This avoids making people afraid of being wrong during discussions. Encouraging a culture of collaborative engagement over judgement and criticism creates an environment for free speech, knowledge sharing, mutual growth and the widespread adoption of BIM across the industry.

This chapter provides an explanation of BIM and its foundational elements, setting the stage for further exploration of key issues in subsequent chapters.

5.2. Disruption driven by clients, not mandates

The last 20 years have seen significant technological changes that have profoundly impacted the operation and transformation of businesses and industries. In this period, the rise of smartphones, apps, machine learning, artificial intelligence and robotics have all disrupted traditional industries.

Starting in 2005, YouTube upended traditional broadcasting by enabling user-generated content. Spotify shook up music streaming in 2006, and Netflix transitioned from DVD rentals to take the lead in video streaming by 2007. Airbnb and Uber disrupted the hotel and transport sectors, respectively, despite facing controversies. Social media platforms, such as Facebook and Twitter (now X), emerged, changing how news and opinions are shared. These innovations have further evolved with the integration of AI, which holds the promise of transforming industries and creating new jobs.

These companies have introduced new ways for consumers to access a multitude of services and enhanced the user experience in numerous ways. Rapid technological evolution is likely to bring further disruption and impact, creating new opportunities for both businesses and clients.

All these disruptions have a common factor: they were adopted by the end user. As stated previously, digital disruption tends to be client-driven, rather than regulation-driven.

Mandates for BIM in the UK

Over the years, two mandates aimed at supporting the digital transformation of the construction industry through BIM implementation have been published. The BIM Mandate, published in 2011 as part of the UK Government Construction Strategy, specified the requirement for fully collaborative 3D BIM by 2016 (Cabinet Office, 2011). More recently, in 2021, the Infrastructure and Projects Authority (IPA) launched its 2030 Roadmap, including an updated Information Management Mandate supporting the UK BIM Framework's implementation (IPA, 2021).

However, these mandates, while aimed at promoting BIM and good information management practices in the construction industry, didn't result in the expected level of adoption. This is often because they aren't client-driven, with some specific exceptions, and there is no enforcement for successful implementation.

For instance, the 2011 BIM Mandate faced challenges in its implementation. Companies struggled to embrace digital adoption as they didn't understand the requirements or benefits. Furthermore, companies weren't held accountable by their clients, for successful implementation, even when stipulated in their appointments. Hence, companies didn't take it seriously as a key project requirement.

I recall the mandate causing uncertainty in the industry, leading to a demand for professionals to support BIM implementation on their projects. However, little progress was made when the mandate was supposed to be implemented in centrally funded Government projects, as there was no accountability for the process's correct implementation or the quality of deliverables.

Following the Scandinavian countries and Austria, the UK, along with many other countries, published BIM Mandates to enhance collaboration and information access. Yet they faced common challenges, such as a lack of standardisation, varying levels of adoption and lack of demand from the end user.

In my opinion, a key reason for the BIM Mandate being less successful than anticipated was the lack of client awareness of the benefits of BIM at each stage of a project, coupled with insufficient education to oversee and support its successful implementation from the start until the end of the project.

The main benefits perceived at the time were improved design coordination and fewer errors on site. However, these were often seen as advantages only for designers and main contractors, despite clearly being beneficial to the client as well. The long-term benefits for the end user, particularly during the operation and maintenance of the asset, were more challenging to convey. This was due to a lack of evidence from other clients and was overshadowed by a culture that favoured immediate short-term results over potential long-term savings.

Just as society didn't require mandates to embrace platforms like YouTube, Spotify, Netflix and Airbnb, the successful implementation of new technologies or methodologies often depends on demand created by a clear understanding of the benefits. These platforms have succeeded because they provide solutions to clients' needs, and the clients, in turn, understand the value they bring. Similarly, for a digital construction industry to truly take off, it's crucial that clients understand how digital construction can fulfil their specific needs. This necessitates raising awareness and educating clients about the advantages and functionalities of the technologies and methodologies in question, fostering a desire for implementation. Generating this kind of client-driven demand can lead to more effective and broad-based adoption, reducing the need for mandates that might not fully achieve their objectives.

By focusing on client education and cultivating this desire, we can shift from an approach dictated by regulations to one that's driven by the genuine needs and understanding of the clients themselves.

Despite this, I believe the BIM Mandates had a positive impact in the industry, as they did raise awareness of the potential of digital technologies to improve the construction industry. They also stimulated the creation of start-ups and inspired technology companies to increase their interest in the construction industry, given the potential that BIM implementation can offer.

Thanks to the UK BIM Framework, BIM implementation is now much easier, providing clearer guidance and processes and less disruptive misinformation than when the first BIM Mandate was published. However, mandates will fail unless they are driven by educated client demand, providing the correct contractual, procurement conditions for success and holding parties accountable for correct implementation. Without this, the value and impact of mandates will be limited.

5.3.   The BIM process at high level

For those who have heard of BIM but are not familiar with it, there is sometimes a perception that it is difficult to implement and fraught with such difficulties that the resulting benefits might not be worth the effort. However, contrary to popular belief, implementing BIM does not require any major innovation; in fact, the implementation process is rather straightforward.

In this chapter, my goal is to debunk common misconceptions about BIM and provide a high-level overview of the process. Specifically, I aim to reassure newcomers that BIM is not an overly complex process. This preliminary overview will help you understand what to expect during the various stages of a BIM project; a more in-depth analysis of the process, as outlined by ISO 19650-2 (BSI, 2021), will be covered in a subsequent sections.

Figure 5.1 High-level overview of BIM process

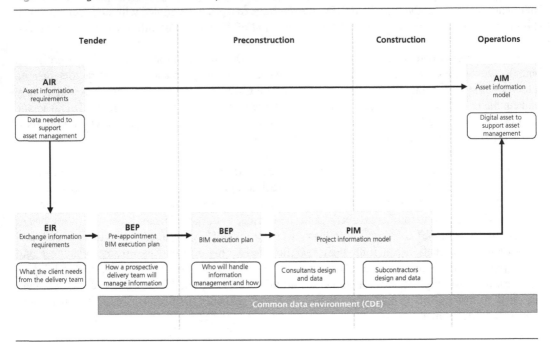

Figure 5.1 has been well received in previous presentations that I have delivered and helps in understanding the process. Conference attendees generally found it easy to grasp the requirements and realised that the process is not as complex as they initially thought.

Tender stage

During this stage, the prospective lead appointed party must review and assess all the information provided as part of the invitation to tender. This includes understanding any details specific to the BIM process, outlined in the exchange information requirements (EIR). Key aspects to understand include what the client expects from the delivery and the kind of information and asset data required to support the business and enable informed decisions. Additionally, the document should specify who is responsible for setting up and managing the common data environment (CDE), as well as outlining the information management requirements for the project's duration.

Every project has its own set of unique considerations, making it essential to fully comprehend all expectations. However, certain factors can affect project delivery, and the expectations and responsibilities are not always clearly defined in the information provided at the tender stage.

As part of the tender response, the lead appointed party will prepare a pre-appointment BIM Execution Plan (BEP) in collaboration with the known task teams, explaining the plan to deliver the BIM-related expectations, as defined in the EIR. Any clarifications regarding the EIR should be sought at this stage.

Preconstruction stage

If the tender stage is successful, the lead appointed party will revise the BIM Execution Plan based on feedback from the appointing party and the participation of any new task teams. Using this approach, the design team will create the design and supply both geometric and alphanumeric information, along with the documentation specified in the BEP.

All project information exchanges will take place within the CDE, adhering to the agreed approval workflow and naming conventions. Importantly, information should not be accepted through email or outside the CDE.

Construction stage

During the construction phase, subcontractors who have design responsibilities or data delivery requirements must adhere to the BEP.

It's important to note that not all task teams will have identical requirements. Some may only need to comply with specific elements of the BEP. Therefore, these considerations should be communicated and factored in during the tender phase and before finalising any appointments.

At this point, given the large volume of data available, consistent management, verification and validation of the data are essential throughout the project. Preparation of the operation and maintenance (O&M) information should not be deferred until the project's completion.

Handover stage

At this stage, all the geometric and alphanumeric information, as well as documentation defined in the BIM Execution Plan, will be transferred to the appointing party. This information, having been validated and verified by the lead appointed party, will accurately represent the completed works. It will assist the appointing party in managing and operating the facility efficiently.

Following my presentation on the high-level process (Figure 5.1), many attendees noted that what we're introducing is not a radical change; rather, these are existing practices, improved and presented in a different format.

Previously, we operated with a client brief and employer requirements. Now, we also have the exchange information requirements, in which clients lay out their expectations for BIM.

Instead of designing in a 2D environment, we have transitioned to working in 3D. This change does come with a learning curve, but it also offers considerable advantages to all stakeholders. One significant alteration is the restriction on sharing information by email. Instead, the requirement is to disseminate it strictly through the common data environment (CDE). This ensures that everyone has access to the latest information and can utilise the appropriate workflows to accept it and distribute it.

Additionally, the information release schedule (IRS) has been replaced by the master information delivery plan, which improves the management of project information.

Lastly, the manner in which we deliver O&M information has evolved. Gone are the days of providing printed manufacturer literature and certificates. We now aim to offer accurate and relevant information, as well as verified structured data, in a digital format. This makes it easier for the facility management team to access the information they need without being overwhelmed by extraneous details.

These adjustments are just some examples that illustrate the differences between traditional methods of project delivery and the new BIM processes and terminology, as shown in Figure 5.2.

Figure 5.2 Before and now (Icons: M.Style/Shutterstock)

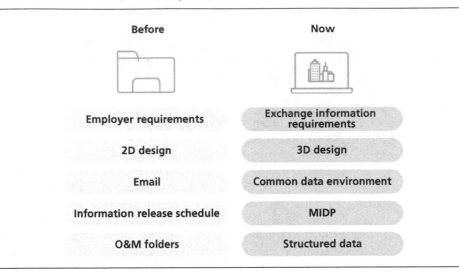

I hope this information has achieved my intention; to explain that BIM and its implementation in a project are not as complicated as they may seem at first. The key to a smooth process is to understand the project requirements, communicate clearly and involve the right teams from the beginning.

We need to acknowledge that this ideal scenario might not always present itself. However, it is exactly these situations we need to transform in our industry if we aim to address such issues as profitability and attracting fresh talent.

Let's not make BIM more complicated than it needs to be. Instead, let's see it as an opportunity to improve design and project delivery and to manage project information to provide accurate and valuable data, meeting the needs of all stakeholders. This cooperative approach will benefit all parties involved, leading to satisfied stakeholders and a healthy bottom line for everyone.

5.4.   Benefits of implementing BIM

Although BIM has gained significant attention in the construction industry since 2011, there remains a widespread misconception regarding its true nature and potential.

The BIM process aims to improve the management and exchange of information, allowing businesses to streamline communication and collaboration among all stakeholders involved in a project. By adopting BIM, businesses strive to enhance the efficiency, reliability and security of information sharing, ensuring that the right information is delivered to the right parties at the appropriate times. This encompasses providing correct and accurate information to support operation and maintenance activities.

When businesses implement BIM effectively, they can unlock the potential of digital construction and leverage a range of digital tools to maximise the benefits of BIM throughout the entire lifecycle of the project.

Now, let's explore some of the benefits that BIM offers at different stages of a project, including design, construction and O&M.

Design and construction phase benefits

- *Improved information management.* The use of appropriate workflows and naming protocols to manage information within the common data environment, accessible to all project participants, ensures that everyone has access to the most up-to-date information, fostering better communication and collaboration.

- *Enhanced decision making.* BIM utilises a 3D design environment that increases coordination and enables the identification of potential errors during the design and construction phase. Clearer visualisation supports the understanding of design proposals and helps all project participants make informed decisions during the design phase.

- *Faster design and redesign.* BIM tools enable quicker production of information and offer an advantage in making rapid changes and updates, compared with traditional methods. By linking design information to sheets and schedules using a few references, the need for manual updates is minimised. This leads to swifter implementation and more accurate representation across all project information.

- *Design for manufacture and assembly (DfMA).* BIM enables the integration of DfMA or modular construction techniques in the design and construction process. By leveraging BIM's capabilities, businesses can optimise the design and construction of prefabricated components or modules. This approach streamlines the construction process, improves quality control, reduces on-site construction time and minimises waste.

- *Utilisation of new technologies.* The outcomes of the BIM process unlock the advantages of incorporating immersive environments, such as virtual reality, augmented reality and construction sequence visualisation. These immersive technologies allow clients to actively engage with the project, leading to a deeper understanding and facilitating informed decision making during the design phase. Furthermore, there is a wide range of emerging technologies that harness the power of BIM to offer additional benefits, including rapid and precise project measurements, advanced data analytics and the utilisation of AI.

- *Reduced errors during construction.* The BIM process supports a clear understanding of the tasks to be undertaken and promotes improved coordination during both the design and construction phases. This substantial reduction in the risk of errors during construction results in a streamlined construction process and minimises any schedule impacts.

- *Conducting appropriate construction verification.* By using point cloud surveys and 360° photography, and comparing on-site progress against the design, the team can identify errors at an early stage, allowing for timely rectification before larger problems occur.

- *Promoting sustainable construction.* By enhancing design coordination and minimising on-site rework, it becomes possible to achieve a significant reduction in on-site waste. This, in turn, positively impacts the construction programme, reduces material costs and decreases the quantity of materials needed on site. These outcomes promote the adoption of more sustainable construction practices.

- *Enhanced safety.* Design models can be used to communicate hazards, both during the design phase and in inductions before site access. This proactively addresses potential risks, helping to ensure a safe working environment.

Operations and maintenance phase benefits

▨ *Complete and verified information.* BIM allows the client to access all essential, complete, validated and verified details through the common data environment. This ensures that the client has reliable current data to support the operational phase, minimising the loss of information throughout various project stages.

▨ *Streamlined handover.* BIM cuts down the time required to give the client the needed information on completing the construction phase. This speeds up the handover process, making sure that the client has all necessary information right from the start to manage and maintain the assets efficiently. This removes the need for extra validation surveys, saving both time and resources.

▨ *Efficient asset maintenance.* During the operational phase, the client's team has easier access to accurate and necessary data, digitally rather than navigation through paper copies. This leads to more efficient asset upkeep by the facility management team. Armed with precise information, the team can proactively tackle maintenance tasks, decreasing the likelihood of asset problems. This reduces the need for repairs and other maintenance activities, ultimately prolonging the asset's useful life and cutting long-term costs.

▨ *Early issue identification.* The available data enable the team to utilise the Internet of Things to identify issues within the assets. This supports more effective maintenance and utilisation, thereby reducing costs throughout the asset's lifecycle.

▨ *Cost reduction during renovation.* Keeping asset information current throughout its lifecycle either eliminates or minimises the costs tied to surveys and inspections during the renovation phase. Accurate data allow the design team to work without the need for intrusive surveys, making the renovation process more cost-effective.

▨ *Regulatory compliance.* BIM ensures that clients have accurate and up-to-date asset information, making it easier to track, trace and locate all assets. This aids in complying with regulatory requirements, ensuring that any necessary updates or changes are made to meet current standards and regulations.

Overall, the BIM process offers effective information and data management, providing a wide range of benefits throughout the entire lifecycle of a building or infrastructure project. By planning the required information from the outset, businesses can improve project management and reduce costs. One of the key advantages of BIM is that it enhances communication, collaboration, decision making and asset maintenance. For a comprehensive understanding of BIM's benefits, refer to Table 5.1, which provides an overview for both the design and construction and the operations and maintenance phases. These benefits will be further discussed in Sections 5.8 and 7.11.

Table 5.1 Checklist: summary of main benefits of BIM (continued on next page)

Design and construction phase benefits
- ☑ Improved information management
- ☑ Enhanced decision making
- ☑ Faster design and redesign
- ☑ Design for manufacture and assembly (DfMA)
- ☑ Utilisation of new technologies
- ☑ Reduced errors during construction
- ☑ Conducting appropriate construction verification
- ☑ Promoting sustainable construction
- ☑ Enhanced safety

Table 5.1 Continued

Operations and maintenance phase benefits
- ☑ Complete and verified information
- ☑ Streamlined handover
- ☑ Efficient asset maintenance
- ☑ Early issue identification
- ☑ Cost reduction during renovation
- ☑ Regulatory compliance

5.5. Challenges to BIM adoption

While this book touches on various challenges and obstacles in revolutionising the construction industry, I'd like to specifically underscore some of the hurdles associated with adopting BIM.

Business support

As previously discussed, any changes can often be met with resistance, especially significant changes that alter the status quo The opportunities afforded by digital construction must not be overshadowed by the fear of change. Therefore, it is critical to have robust backing from high-level decision makers, such as senior leaders and directors, to encourage acceptance and cooperation throughout the organisation. They play a vital role in promoting the acceptance and execution of new processes across all business units.

The BIM process affects everyone's roles; this means that support is needed from all levels within the organisation. Team members must each understand their roles in this process and take ownership of their required actions. It requires a culture of accountability, facilitated by top management, to ensure the successful implementation within the business.

Misconceptions and unsuitable partners

One common misconception about implementing BIM in a project is that it will increase costs and delay the schedule. This misconception is often due to a lack of knowledge and past experiences of working with unsuitable partners who might not be fully equipped to handle the demands of BIM. Additionally, some team members might feel intimidated by the introduction of new processes and technologies instead of embracing them and undergoing the necessary training to improve productivity. Adequate training and communication are needed to dispel these misconceptions within your organisation.

As highlighted in Chapter 3, the success of BIM implementation in your projects largely hinges on the competency of your partners. The saying in the construction industry that 'you're only as good as your supply chain' holds true here. Therefore, involving well-equipped, knowledgeable partners in your project is the foundation for successful BIM implementation.

Mindset barriers and disbelief in the benefits

The construction industry is often resistant to change and hesitant to embrace new technologies. Many professionals hold onto the mentality of 'I've always done it this way,' which can hinder the adoption of BIM and digital construction. Even clients may be reluctant to adopt BIM, owing to negative perceptions or fears of increased costs, despite being aware of the potential benefits.

It is crucial for both professionals and clients to recognise the value and benefits of adopting the BIM process to stay competitive and reap the advantages it offers. The most effective way to address resistance and disbelief regarding the benefits of BIM is by sharing feedback and stories from colleagues involved in BIM projects.

Lack of proper planning

There is often pressure in construction projects to start work on site as quickly as possible, even if the design is incomplete or unverified. The belief is that starting early will expedite project completion and allow for earlier billing. However, it is important to approach projects with patience and proper planning, ensuring a fully verified and robust design before commencing on-site work. Rushing can lead to problems on site, resulting in costly delays and rework.

Neglecting long-term considerations

In many projects, the design and construction phase is prioritised, neglecting the long-term legacy and maintenance aspects. It is crucial for construction professionals to not only focus on the quality of finishes but also ensure that the correct specifications, information and data are provided to support the operations and maintenance phase of the project. Considering the client's needs beyond project completion enhances overall project success and client satisfaction.

Communication and management of information

Successful project delivery depends on effective communication and information management. Poor communication and information loss often lead to errors and rework on site because team members lack access to the latest accepted information. In an attempt to meet tight deadlines, there is often a disregard for established standards, relying on faster but riskier communication methods, such as email, as well as the issuing of comments on a printed copy rather than in a digital format, which lack security and proper traceability.

Additionally, the isolation of design and delivery teams from the end user can lead to the needs of the end user being overlooked or the omission of relevant information for effective operations of the facility. To address this, involving the facility management team early on in the project is crucial. Their involvement ensures that data are collected according to the appropriate exchange information requirements in the correct format, and any concerns or queries regarding information handover can be resolved effectively.

Technology adoption

Many companies view technology as an expense rather than an investment. However, with the correct process, technology can minimise errors during the design and construction phases, provide additional services to clients, differentiate businesses from competitors and enhance the overall client experience.

Today, technology advances quickly and business needs to acknowledge the importance of adapting to stay ahead of the competition, standardising processes and investing in the right tools. By adopting technology with a well-defined strategy and process, and choosing the most fitting tools, companies can experience a swift return on their investment.

While the benefits of adopting digital construction are significant, it's clear that there is also considerable resistance and various challenges to overcome. Table 5.2 summarises some of the common challenges and misconceptions that you'll need to address during the implementation of digital construction within your business.

Table 5.2 Checklist: challenges to adopting BIM

Business support

☑ Conduct change management workshops to ease team members into new changes.

☑ Secure endorsement and active support from senior leaders and directors.

☑ Roll out comprehensive training to clarify everyone's role in new processes.

Misconceptions and unsuitable partners

☑ Dispel myths through educational materials that demonstrate how BIM can improve project delivery, saving costs and time.

☑ Vet partners for BIM compatibility before entering into collaborations.

☑ Offer support and learning courses to familiarise team members with new technologies and processes.

Mindset barriers and disbelief in benefits

☑ Implement campaigns to attract supporters and influencers.

☑ Organise workshops to communicate to teams how digital construction will benefit their day-to-day jobs.

☑ Provide case studies to customers that demonstrate the benefits of adopting digital construction.

Lack of proper planning

☑ Establish the need to have a complete design before commencing on-site work.

☑ Emphasise the necessity of a fully verified and robust design through internal audits.

Neglecting long-term considerations

☑ Incorporate long-term maintenance and operations into the planning and design phase.

☑ Communicate the importance of providing accurate data to meet the needs of the appointing party during operations and maintenance.

Communication and management of information

☑ Adopt effective information management processes to prevent data loss.

☑ Facilitate regular meetings between delivery teams and the appointing party.

Technology adoption

☑ Conduct a cost–benefit analysis to showcase technology as an investment based on the business needs.

☑ Develop a comprehensive strategy for the adoption and deployment of new technology.

5.6. Keep it simple

Digital construction is a very intriguing and stimulating subject. Therefore, it is crucial to avoid rendering it tedious and dull. Our aim is to keep it engaging so as to draw people in, rather than deter them.

One of our primary objectives is to simplify processes, especially if we desire a broader acceptance and adoption of BIM within the industry. A straightforward approach achieves three pivotal outcomes. First, it eradicates potential hindrances and bottlenecks, facilitating seamless and efficient teamwork. Second, the removal of distractions enables core objectives, such as information management, design coordination and project management, to be directed towards truly meaningful goals, instead of being bogged down in needless complexities. Lastly, uncomplicated BIM guidelines encourage adherence to industry standards, which in turn minimises errors. When professionals are provided with clear and straightforward guidelines, the likelihood of making mistakes significantly decreases.

This section aims to underline the substantial advantages of straightforward BIM processes and offer assistance in refining the design and construction process, keeping it engaging and inviting to all stakeholders.

The shift to information management

Over the years, the term BIM has unfortunately gained a negative reputation for various reasons. These include poor implementation, a lack of skills among construction industry professionals and inflated fees from some companies for delivering BIM requirements. However, one of the most significant challenges has been achieving a common understanding of BIM among industry professionals. This has led to the widespread belief that BIM is primarily a 3D design tool that is expensive to implement and takes longer to use to deliver projects, despite the potential benefits. The absence of a standardised approach and clear communication about BIM's actual requirements and benefits has further hindered its broad acceptance across the industry.

Compounding these issues is the emergence of an entire industry around BIM, where some actors have unintentionally complicated the understanding of the process and requirements while selling their services.

Recently, efforts have been made to shift the focus away from BIM and towards the concept of information management. This change in focus is an attempt to address the common misconception that many people have when they hear the term BIM, that it is often limited to 3D modelling. By emphasising the importance of the process and of information management throughout the entire lifecycle of a building, it is hoped to convey the true value and potential of BIM and to promote its adoption for the benefit of the construction industry and the businesses involved in a project.

I have always conceptualised BIM as a process for managing information throughout the entire life-cycle of a building, and the use of 3D models as an improved way to produce and coordinate design information, with the ultimate goal of delivering a better product and information to the client. It improves collaboration, communication and decision making, reduces errors and ultimately leads to better built and more efficiently managed assets. However, I have personally avoided using the acronym BIM in recent years, not because I believe it is incorrect, but because many people tend to become defensive or hesitant when they hear it, probably because of a lack of understanding of the concept.

I agree that there is a clear need to separate the association that has developed over time between BIM and 3D models. However, it is important to note that BIM is a concept that has been around for decades; while it may seem that moving away from the term BIM and focusing on information management is a simple solution, this can also lead to confusion. In addition, the industry has been adapting to changes caused by the switch from PAS 1192 (BSI, 2013), BS 1192 (BSI, 2016), and ISO 19650-2 (BSI, 2021) to manage information through the years. During this transition, education and onboarding are required; too many changes or updates in a short time could lead to a lack of confidence and doubt in the process.

The key to overcoming the challenges associated with BIM lies in raising awareness and fostering a more comprehensive understanding of the concept throughout the industry. This can be achieved through better education, standardisation and simplification of the process. It is important for stakeholders to recognise that BIM is not just about 3D modelling or software tools but is an all-encompassing approach to managing information and collaboration throughout the entire lifecycle of a building project.

Fundamentals of BIM implementation

Incorporating BIM in projects can be straightforward, provided we don't overcomplicate the process, which might discourage widespread adoption across the industry. Fundamentally, BIM is about managing information effectively, an aspect often overshadowed by the focus on 3D design. However, without following some basic principles, even the most impressive and detailed 3D models will fail to add value.

There are three critical principles for effective information management in any project that your business tackles. It is important to keep these principles in mind throughout any project, as they lay the groundwork for successful BIM integration.

These principles will be explored further in subsequent sections of this book. They cover:

- Common data environment.
- Naming protocol.
- Workflow for accepting information.

While other elements, such as design model coordination, asset data, the security approach or health and safety or fire safety requirements will define the BIM implementation, it is when these requirements are overwhelming or not communicated in a clear and concise manner to the teams that resistance and pushback to its implementation occur.

The three principles of proper information management form the foundation for smooth information exchange and enhance efficiency, accuracy and timeliness in utilising project information. By adhering to these principles, teams can boost confidence in the process and improve the quality of project outcomes.

Emphasising simplicity in BIM

Embracing simplicity is a fundamental principle that should be applied across all aspects of a project, including clients' documentation and communication. Often, clients submit extensive documents with a wealth of asset data requirements during the tender stage. These documents are typically based on generic templates created by external consultants without considering the specific needs of the project or the client's internal strategy. However, advocating for simplicity does not mean that clients should lower their expectations. Instead, it is crucial for them to have a clear understanding of what they want, why they want it and how they intend to utilise the information requested to ensure it delivers value.

For example, requesting a large amount of asset data as part of the project requirements within the BIM documentation at the tender stage, when the client does not have a basic strategy to manage this information and does not know what to do with it, and when the client's needs are a bare minimum to support facility management, will not benefit anyone. Regardless of how, in theory, all this asset data could be useful, in practice, it will have the opposite effect.

By focusing on simplicity, clients can effectively communicate their expectations to potential lead parties during the tender stage. If project teams find it difficult to comprehend the information presented, they may be less inclined to adopt the proposed approach. This reluctance could lead to additional fees to cover the risks associated with the unknown. Therefore, maintaining simplicity and transparently conveying the client's needs are essential to prevent confusion, additional costs and potential misunderstandings.

The aim of this simplicity approach is to proactively prevent disputes that often arise from unclear requirements, misunderstandings, or overlooked information. By gaining a thorough and clear understanding of clients' specific needs and expectations, the lead appointed party becomes more capable and equipped to communicate and disseminate the requirements with the different task teams involved in the project and navigate the challenges associated with BIM adoption. A persistent commitment to this straightforward and transparent approach not only ensures smoother project execution but also substantially minimises the risk of disputes, ultimately benefiting every stakeholder involved.

Each project has unique characteristics and requirements, and it is essential to acknowledge challenges and engage in open discussions with the team about any necessary minor deviations or clarifications from ISO 19650 before commencing, if required. The ISO 19650 standard allows for minor adjustments, but it's fundamental to ensure that any necessary deviations are agreed on and recorded in the BIM Execution Plan. By keeping the process straightforward and ensuring that all deviations are recorded in the plan, project teams can ensure smoother collaboration with all parties and more efficient project delivery while maintaining compliance with ISO 19650.

To successfully promote the implementation of BIM, simplicity is of utmost importance. Overcomplicating the process by requesting excessive data, asset tags or overwhelming workflows that do not serve a genuine purpose for both the clients and the project team can be counterproductive.

Enhance communication and streamline documentation

A common issue in many projects is the excessive use of outdated templates, by the appointing party team, that do not meet the project's specific needs or expectations. This can give the impression that the appointing party is not genuinely committed to the project's BIM success. While templates can be a useful starting point, they should be tailored to meet each project's unique requirements. By adapting templates to each project, teams perceive the client's commitment to a positive outcome of the BIM process and the use of deliverables after the handover stage.

Another frequent problem with BIM documentation is the prioritising of content quantity over quality, leading to lengthy documents that are difficult to understand and potentially confusing for project stakeholders. The primary goal of BIM documentation is to provide clear and accurate information that supports a building's design, construction and operations. Therefore, it is essential to ensure high-quality, relevant content with realistic expectations. Project teams should concentrate on producing concise, clear documents directly related to the project's goals. Lengthy and redundant documentation only creates confusion within teams, hindering BIM adoption in the project.

Moreover, the appointing party should actively supervise BIM implementation and be available to address any questions or concerns about documentation content. This helps maintain the project's progress and enables the prompt resolution of any emerging issues. Active involvement of clients ensures that the BIM process matches their goals and expectations, ultimately contributing to a more successful project outcome.

To enhance clarity and simplify, it is essential to eliminate redundancy between the documents related to traditional processes and the new ISO 19650-compliant documentation. This helps prevent teams from feeling burdened with extra or duplicated tasks. As companies transition to BIM, some old processes and documents can be replaced with new ones that comply with ISO 19650.

However, there are instances where both sets of documents continue to be used, and this issue needs to be addressed. Examples include the MIDP, TIDP, responsibility matrix and level of information need. The content of these documents is similar to that of the traditional documentation commonly used in the industry, with which teams are already familiar, such as the information release schedule, design delivery programme, design responsibility matrix and deliverables matrix, as part of the scope of services. The documentation required by ISO 19650 should be viewed not as an added burden but as enhancing traditional documentation, offering better information to support project delivery and clarify roles and responsibilities. To prevent confusion and redundancy, which can hinder project expectations and cause conflicts between documents, companies should evaluate their internal processes and documentation. By ensuring that clear and concise information is provided, companies can better support the understanding of project expectations.

Also, when communicating and documenting the BIM process, it is crucial to use clear and understandable language. Jargon and acronyms can be confusing and off-putting for some team members, making them less inclined to learn about the process. It is essential to use language that is accessible to all team members, regardless of their BIM expertise.

In summary, effective BIM implementation depends on streamlined documentation and clear communication of the process. Emphasise simplicity by eliminating redundancy, prioritising quality content and tailoring documents to project needs. Engage the project team using accessible language and delineate responsibilities for efficient project execution. No one desires to embrace complexity; by addressing these aspects, companies can optimise BIM processes for improved project outcomes.

5.7. Information management

Information management principles

As I previously pointed out, in my view, the BIM process is straightforward and revolves around three core principles.

- The first is the *common data environment* (CDE), which provides a centralised location for all project information. It's essential that the CDE uses appropriate metadata to enable easy retrieval and tracking of information.

- Second, the *information approval workflow*. A well-designed workflow should be both effective and efficient, ensuring that all project information is distributed to relevant parties for comments and approval. This workflow drives the success of BIM implementation and helps to guarantee that all information is correctly managed throughout the project.

- Lastly, the *naming protocol* is crucial to ensure that information generated during the project adheres to naming conventions agreed with the client, and that compliance is verified as part of the project workflow. Implementing the proper naming protocol not only meets the client's expectations but also assists all project stakeholders in locating information within the common data environment more efficiently.

Before diving into the details, it's important to note that the principles of information management are not exclusive to BIM projects. They are universally applicable across various types of project, whether in the building sector, civil engineering or other industries. Effective use of a common data environment, adherence to approval workflows and strict compliance with naming conventions are beneficial practices for any project. These principles aim to streamline information flow, improve collaboration and enhance overall project efficiency, regardless of the specific sector you are operating in.

Although I believe these principles are straightforward, it's important to acknowledge that their practical application can be quite challenging. It's not uncommon for team members to struggle to fully grasp their significance or to recognise the substantial potential benefits they can offer.

Therefore, the success of BIM implementation relies heavily on cultivating a work culture that encourages collaboration, communication and an eagerness to learn and adapt to new technologies and methodologies. To clarify, while the appropriate technology can enhance and streamline the BIM process, it's ultimately the people involved who drive its success, regardless of the project's size or complexity. When all participants understand and commit to their respective roles in the BIM process, adherence to these principles not only optimises project delivery but also provides an added layer of protection for your business.

Those involved bear the responsibility to ensure that processes are followed, workflows are developed and executed, communication and collaboration are maintained and technology is effectively utilised to achieve the desired outcomes.

It is true that technology plays a central role in the digital transformation of the construction industry, but while it cannot be ignored, it should not be the primary focus when integrating BIM in your business. Instead, the right technology should complement and enhance established, standardised business processes and facilitate effective information management.

The common data environment

According to BS EN ISO 19650-1 (BSI, 2019), a common data environment (CDE) is defined as 'an agreed source of information for any given project or asset, for collecting, managing and disseminating each information container through a managed process.'

Further, BS EN ISO 19650-2 (BSI, 2021) outlines the required functionalities of a project's common data environment, which plays a crucial role in the success of the project by serving as a centralised location for all project information, from tender to handover and during the operations and maintenance phase. This ensures secure data management and facilitates structured dissemination of project information through agreed naming conventions and approval workflows. This streamlined process enables easy retrieval and tracking of data.

By utilising the CDE with the appropriate workflows and naming protocol, project participants can access the most up-to-date information and collaborate using it in a safe and secure environment. This provides consistency, accuracy and currency to project information, reducing the risk of information loss or misuse, and maintaining a record of comments and acceptance records that might be needed in the future in case of disputes.

There are many advantages to adopting a single source of information using the correct implementation of a CDE.

- *Preventing loss of information.* By centralising all project information in the CDE, the risk of losing important information communicated through various methods, such as emails, is eliminated. This ensures that all information is stored in one secure location and is easily accessible to the relevant individuals.
- *Peace of mind.* Having a single source of information allows project participants to work with the latest and most up-to-date information available, promoting clarity and coordination. This provides assurance that all project information is always available to the relevant parties.

▨ *Easy access to information.* Productivity is boosted as the right information can be found quickly, saving valuable time that would otherwise be spent searching through various folders.

▨ *Audit trail and tracking.* The CDE maintains a clear audit trail of all changes made to any documentation, including information about who made what changes and when. This facilitates easy retrieval of information when needed and ensures a comprehensive record of all comments and acceptances, which can be referenced in the future if necessary.

▨ *Better coordination.* By using the latest accepted information from the CDE, team members can work more efficiently and reduce the number of on-site issues. This ensures that everyone is working with the same set of information, eliminating confusion or misunderstandings and avoiding rework caused by using incorrect information.

▨ *Secure access groups.* The CDE allows for the creation of access groups for specific information, enabling compliance with security requirements. This helps protect sensitive information and maintain confidentiality.

▨ *Reporting and finding information.* The CDE provides a clear overview of information status and revisions made and facilitates quick information retrieval. This allows for more efficient reporting, identifies information that hasn't been revised on time and aids in making informed decisions based on accurate and up-to-date information.

Strategies for implementing a common data environment

When implementing the common data environment (CDE), two situations tend to arise.

PROJECTS WHERE THE APPOINTING PARTY LACKS A CDE TO MANAGE PROJECT INFORMATION

In instances where the appointing party doesn't have a CDE to manage the project information, it is common that the lead appointed party is appointed to take responsibility for implementing, configuring and supporting the management of the CDE.

On project completion, the appointing party can then receive the information for their records and transfer it to their computer-aided facility management (CAFM) system. This enables effective asset management during the subsequent operations and maintenance phase.

In accordance with standard recommendations, it is advised that this implementation be handled as a separate appointment. This appointment should be completed before the procurement of any other appointed party begins, and the project CDE should be in place prior to issuing the invitation to tender. This ensures that information can be shared with tendering organisations in a secure manner. It's worth noting that this task could also be performed by a third party appointed by the appointing party if preferred.

PROJECTS WHERE THE APPOINTING PARTY HAS A CDE AND TAKES RESPONSIBILITY FOR INFORMATION MANAGEMENT

In projects where the appointing party has a CDE to manage the information, the lead appointed party may still choose to establish a separate CDE.

This arrangement allows the lead appointed party to manage and track information from different task teams effectively. Such a setup facilitates better tracking and tracing of information, ensuring that it can be retrieved in the future if necessary.

The information is only transferred to the appointing party's CDE once it has been authorised by the lead appointed party and accepted by the appointing party.

In this scenario, clarifying the information transfer process is vital. This includes understanding how information will be transferred to the appointing party's platform during different stages of the project. Some appointing parties may prefer partial transfers at various stages of the project, while others may opt for a full transfer at the end.

This process must be communicated and agreed early in the project to avoid confusion and ensure a smooth transition of information. Careful coordination ensures that all parties have the information they need, when they need it, maintaining the integrity and accessibility of the project data throughout the project's lifecycle.

Managing information with a security-minded approach

For the successful execution of a CDE, it's important to understand the specific requirements of the appointing party. Whether you're setting up a new CDE for a project or working within an existing one, you must be aware of these requirements. This allows you to plan the exchange of information effectively, ensuring that all data are properly managed, transferred and archived for all parties involved.

Exchange of information with a security-minded approach includes the following.

- *Exclusivity in data sharing.* All project data must be shared exclusively through the common data environment (CDE), in compliance with established security criteria from the start. Circulating information by email is not only inefficient but also risks data loss and security breaches for stakeholders. Any deviation from this protocol is strictly prohibited and should be penalised.
- *Collaboration is necessary.* Effective collaboration and adherence to established processes make it easier to track information, thereby reducing the likelihood of errors or delays.
- *Team access control.* Remove team members from the CDE who have left the project, and again on project completion, to ensure ongoing data confidentiality and project safety.
- *Sensitive information protocols.* Recognise classifications of sensitive data and implement various approval methods for added protection. Use the MIDP to identify what information requires restricted access and develop a clear management strategy during both the design and construction phases.
- *Adherence to appointing party security protocols.* When working with an appointing party that has its own security procedures, it's crucial to use a CDE that aligns with those requirements to maintain data confidentiality and integrity. An assessment of the CDE and its suitability will be necessary. This process can be time-consuming, so it's important that the appointing party communicates these requirements as part of the tender documentation, enabling any required investigations to begin well in advance.

Especially in projects with unique security requirements, it's essential to establish a well-defined strategy for information dissemination. Experience from previous projects has shown that unclear or convoluted security requirements can hinder the correct exchange of information and delay access to the necessary details for progressing the work according to the schedule. Therefore, upfront discussions with the appointing party to clarify these strategies are invaluable to minimise any negative impact on the project.

Furthermore, in projects that require security clearance, it's best practice for stakeholders to complete these clearances well in advance. The clearance process can occasionally be lengthy, and individuals will not be able to work on the project until it is completed. Therefore, taking care of this promptly ensures that administrative tasks won't cause delays once the project is underway.

The structured framework offered in ISO 19650-5 (BSI, 2020) helps organisations understand potential vulnerabilities. It outlines the controls needed to manage security risks without compromising collaborative efforts.

For a more in-depth exploration of prioritising security, the National Protective Security Authority (2023) offers comprehensive guidelines for adopting a security-minded approach to information management.

Building a comprehensive security management plan

Businesses need to develop a security management plan, led by a security specialist, which serves as a roadmap for strong cyber protection. This plan also reassures clients about data security. To build a robust information security management system, the following elements are to be considered.

- *Industry certifications.* Regular security training and workshops for staff are a must. Such certifications as Cyber Essentials Plus (National Cyber Security Centre, 2023) and ISO 27001:2022 (ISO, 2022) demonstrate a strong commitment to security.

- *Proactive strategy and management.* Businesses should focus on prevention rather than reaction. This means creating a culture that values security from the start and having a team that actively manages security every day.

- *Clear communication and data protocols.* Normal communication channels are fine for general data, but sensitive information requires encrypted routes. A secure CDE with tight access controls is key for storing crucial data.

- *Risk management.* Businesses should identify threats and vulnerabilities systematically, and take such steps as data encryption and secure data transfer to prevent risks.

- *Backup and information handling protocols.* A mix of on-site and cloud backup solutions is important. Regular recovery tests and compliance with government and industry classification standards make information management more efficient.

- *Robust access, security training and password policies.* Careful monitoring of third-party access, ongoing security training and strong password rules strengthen a company's defences.

- *Transparent incident reporting and secure destruction.* It's important to create an environment where staff can report security issues without fear. Also, there should be guidelines for the safe disposal of sensitive materials.

- *Business continuity plans.* Alongside cloud solutions, businesses should have a detailed plan for continuity and recovery.

By thoughtfully adopting and consistently maintaining these security measures, businesses can strengthen their defences and build lasting trust with stakeholders.

The importance of the information manager's role

While the advantages of using the CDE are clear both for the project and the business, some users may initially resist this change. This reluctance often arises from a departure from the familiar process of sending information by email and the learning curve associated with using the CDE instead of email or a local drive. Furthermore, the CDE provides a transparency that reveals each individual's actions, which is a level of openness that the construction industry isn't accustomed to.

This resistance underscores the importance of the information manager, whose role is key in ensuring that the CDE is successfully implemented and information is correctly managed in projects. This entails offering the necessary support and training to all teams, and overseeing correct application of the CDE throughout the project.

It's essential to have a clear and standardised approval process, both internally and externally. All team members within an organisation should adhere to the same approach across different offices, enabling the information management team to implement the same workflow with minor deviations based on project requirements. Establishing this standardised approval workflow requires consensus from the design management and operations team. While specifics might vary based on your company's roles, I recommend adopting a unified approach and securing the acceptance of teams contributing to the CDE, easing new project implementations through familiar processes..

The CDE extends beyond being just a tool for the project. Properly managing the CDE is imperative; it acts as a protective measure for businesses. It ensures that all data, communications and actions are logged, tracked and easily retrievable, supporting the project lifecycle's success and safeguarding your business in the event of disputes.

Despite its significance, proper information management is sometimes overlooked in the industry. Nonetheless, owing to increased regulations and industry efforts to maintain a clear information chain, companies are becoming more attuned to its importance. While the industry has seen progress over the years, more effort is needed to fully utilise the CDE and ensure timely completion of tasks, moving away from sharing and commenting on drawings outside the CDE.

When introducing the CDE in a new project, all pertinent team members should access it promptly. Everyone must receive training, even if they are familiar with the platform, to comprehend the latest best practices for information sharing and review. Clear roles must be established, specifying the primary contact responsible for reviewing information at each workflow stage and setting action timelines, ensuring quick information dissemination through the CDE.

While deployment of the CDE is vital, it's equally important that all uploaded data meet the set guidelines. This can be achieved only through consistent monitoring and evaluation of the information. I advocate for weekly project reports as a cornerstone of information management. These reports capture and share the status of all issued information, ensuring accountability and transparency. To generate these reports, data must be extracted from the CDE, offering summaries that highlight:

- information pending review and the parties with outstanding comments, spotlighting potential delays;
- a discipline-wise project status, showing the percentage of accepted or rejected information, helping the team gauge the project's progress.

Such reports not only inform teams about pending CDE actions but also assess the platform's information quality. They hold stakeholders accountable for addressing outstanding tasks efficiently. By regularly producing and sharing these reports, the team can clearly understand the project's progress and make informed decisions.

While the reports generated by the information manager are invaluable, they become futile if not acted on. It is negligent for the team not to address highlighted actions. These reports should be

Figure 5.3 Information states according to ISO 19650-1 (adapted from BSI, 2019)

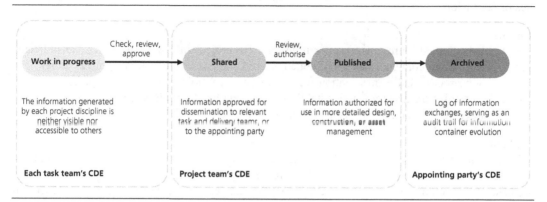

integral to the senior team's discussions on project progress. When generating reports based on the data available within the CDE, their outcomes must enable thorough analysis, ensuring relevance and providing the necessary information for the team to make informed decisions throughout the project. It's important to avoid producing voluminous reports lacking a defined purpose, as this can introduce undue complexity and result in wasted time, with teams likely disregarding such reports.

Information states

As each item of information produced for the project evolves over time, it goes through different states, as described in Figure 5.3, based on ISO 19650-1 (BSI, 2019).

WORK IN PROGRESS

As specified in the guidelines of ISO 19650-1, the 'work in progress' (WIP) state pertains to the initial stages of information generation. During this phase, individual task teams develop and refine the necessary project information. Crucially, information container in the WIP state should be held exclusively by the originating task team.

This is a designation for unverified design data that haven't yet undergone broader scrutiny – essentially, raw, evolving data that the internal design team use as they work towards a more polished, shareable version. Although it's technically possible to store this information within the project's common data environment (CDE), I personally haven't used the project CDE for such information. This is because design teams usually have their own platforms to share work-in-progress information, either within their internal task team or collaboratively with other specific external task teams. If you decide to use the project CDE for this purpose, it's crucial that you manage the access permissions correctly and ensure that WIP isn't accessible to other teams or external stakeholders.

SHARED

Once information advances beyond the WIP stage, it transitions into the shared state. This state represents information that has undergone the check, review or approval process. The shared state within the CDE is a collaborative hub where all project stakeholders can access the latest refined information. This facilitates cross-team collaboration, allowing stakeholders to review, comment on and collaboratively work with the information. To ensure accountability and a clear audit trail, information in the shared state should be immutable once logged into the system, until it is accepted or rejected, This will then allow for a new revision. This approach guarantees that all teams reference a singular, consistent version of the information, eliminating potential discrepancies or confusion.

PUBLISHED

The 'published' state represents a major milestone for project information. This state indicates that the information container has been refined, shared and passed through a rigorous review and authorisation process. Once information reaches this state, it holds contractual significance. This essentially certifies the accuracy and suitability of the data for the designated purpose. Stakeholders can, therefore, rely on this published information for design and construction

ARCHIVED

All information that has been shared and published in the CDE throughout the project duration is preserved for knowledge retention and for regulatory and legal purposes. My recommended organisational method for the CDE is to include metadata based on the Royal Institute of British Architects (RIBA) project stages. Categorising the archived information based on these RIBA stages facilitates swift location of specific information and offers a chronological view of project progress. This structured approach also guarantees the information's availability for future reference, especially for postproject evaluations and ongoing maintenance. During the project's handover phase, pertinent archived information can be transitioned to the client's CDE, adhering to the stipulations agreed with the client. This handover information provides to the client with a comprehensive overview of the project, supporting an efficient postcompletion management.

Additionally, in Figure 5.3, I have illustrated a common approach from the perspective of the lead appointed party. This approach is predominant in most projects, where different task teams work within their own common data environment (CDE) during the production of information. To recap, it is noteworthy that while the standard states that the work in progress information should be neither visible nor accessible to others, some task teams might allow other task teams to access their CDE, to collaborate on the production of information at this stage, before it is shared with the wider delivery team. However, it's important to exercise caution when using work-in-progress information.

Once the information has been checked, reviewed and approved by each task team, it will be shared within the project CDE. This environment, is normally is set up and managed by the lead appointed party, where the rest of the project team can review the information container, and it will be authorised and accepted for use.

Following this, the lead appointed party will archive the information according to their own needs and records. The relevant information requested by the appointing party will then be transferred to the appointing party's CDE, following the agreed process.

As mentioned earlier, this approach may vary based on the project needs, the solutions implemented and whether the appointing party has their own CDE. However, the described approach is the one I have come across most often.

Workflows

A smooth and consistent flow of information is fundamental in any project. The uploading of data to the common data environment should meet the project's established requirements and the data should be quickly and easily accessible to all relevant stakeholders. This isn't simply about moving data around; there's a methodical process for transitioning information, guided by set authorisation and acceptance processes.

Checks, reviews and approvals transition

As mentioned earlier, the work-in-progress information resides solely within the task team's own systems, whether that's an internal network or a cloud platform. It's only when the information successfully navigates the initial gateway review, which is conducted outside the project's CDE, that it becomes eligible for wider dissemination.

Before information is shared with a wider audience within the project CDE, it undergoes thorough checks, reviews and approvals by the task team. This entails cross-referencing the information against an established information delivery plan and ensuring that it aligns with the agreed standards and methodologies. Only after the task team lead gives approval is the information uploaded to the CDE.

Review–authorise transition

Next, every item of shared information undergoes a review to ensure that it is fit for purpose. If it meets the requisite standards, its status updates to 'published'. If not, it's sent back for further refinement. It's crucial to understand that the nature and purpose of the information dictate its workflow, with different purposes sometimes requiring distinct processes.

Aligning the workflow with your overall business strategy is essential. Your business should clearly define which team members are responsible for reviewing and commenting on the information provided by each task team, as well as identifying any other parties required to review the given information. Moreover, understanding the client's preferred level of involvement is crucial. Such alignment ensures that all parties meet their objectives. It's also important to be attuned to the client's needs and preferences; some may want to be actively engaged in the process, while others might lean towards a more passive role.

Managing information in projects varies based on the intended purpose of the uploaded information within the common data environment. The choice of workflow should match your standardised approach and any specific project requirements should be factored in.

For example, as illustrated in Figure 5.4, information that is shared exclusively for informational purposes using the code S2, as described in Table 5.4, has a distinct workflow. The information management team is responsible for ensuring that the content adheres to the project's established protocols. Specifically, the team make compliance checks to ensure that the information container uses the correct naming conventions, metadata and title block in the document, in accordance with the MIDP. If the uploaded content meets these standards, the information management team

Figure 5.4 Information workflow: S2, information

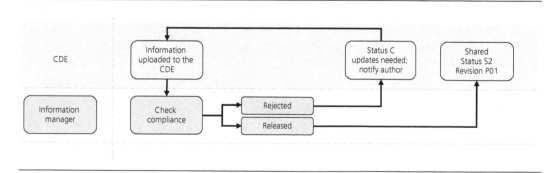

Figure 5.5 Information workflow: S3, review and comment

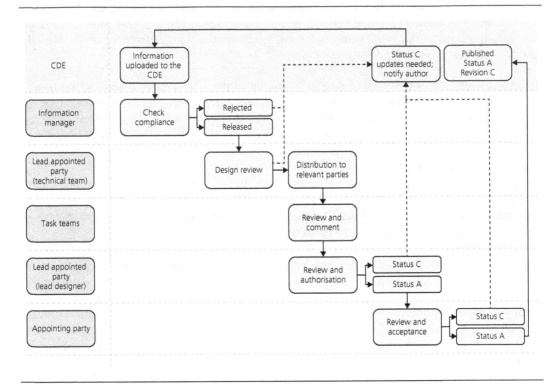

releases it, making it visible and accessible to the entire project team. Should the content fall short of these criteria, it is sent back for further revisions. The process for S2 information is deemed complete once the information manager has validated its compliance.

Conversely, Figure 5.5 shows the workflow for information intended for reviews and comments. This uses the S3 code and has a lengthier journey, as depicted in the figure.

In this scenario, when information is shared for review and comment, the initial compliance checks are made in the same way as when information is shared solely for informational purposes.

After the information has satisfactorily passed the initial compliance checks, it is reviewed by the lead appointed party's internal technical team – which comprises design managers and technical services managers – If the review is satisfactory, the information is then distributed to the appropriate task teams. However, if the team deems the content non-compliant with the scope and the project's information production methods and procedures, it will be rejected with appropriate feedback provided for the task team to make necessary amendments.

The objective of this step is to collect the technical input from the relevant task teams and ensure they are aware of and satisfied with the information.

This approach also aims to prevent overwhelming various teams with unnecessary information, ensuring only relevant parties have active involvement. There's an intention to filter the information, with the goal of sparing teams from reviewing non-relevant data. If someone introduces content outside the agreed scope, it's the responsibility of the lead appointed party to reject this without wasting the team's time.

The technical input gathered from these parties is made available to the originator and visible to the whole project team. The necessary modifications will be incorporated in the revised information.

The information management team will report on the comments received and any missing items. The technical design team will then take the necessary actions, as some comments concerning specific information may be more urgent than others. The lead appointed party is responsible for reviewing and coordinating the comments provided by the different task teams, ensuring that all necessary comments are received from all relevant parties.

Based on the technical input provided by different task teams, the lead appointed party determines whether the information corresponds to the project's information production methods and procedures.

The information is then either authorised or rejected, being assigned Status A or Status C. While the lead appointed party is primarily responsible for this task, the agreement between the lead appointed party and the lead designer usually specifies that the lead designer will take on the responsibility of coordinating and defining the appropriate status of the information.

Please be advised that Status B, defined as 'information containers that have been partially signed off, with comments from the invited authorising or accepting party', is marked as deprecated in BS EN ISO 19650-2:2018 (BSI, 2021). Therefore, it is recommended to use only Status A or Status C.

If the content is found to be unsatisfactory, Status C is awarded, indicating rejection. Necessary adjustments are then made based on the technical input and a new revision is issued to the CDE. It's important to note that teams must review the rejected information, based on the comments, within the agreed time frame. If the information is no longer required and needs to be withdrawn, this should be clearly stated within the CDE, and the master information delivery plan (MIDP) should be updated accordingly.

Once the lead appointed party has authorised the information container, this will be shared with the appointing party for review and acceptance to become a contractual document. However, if the appointing party is not satisfied with the content, they retain the right to reject it. In such cases, the workflow would recommence with the revised information, ensuring that the comments have been addressed. Please note that some appointing parties may not wish to be involved in the acceptance process, and this responsibility will fall to the lead appointed party. Therefore, it's important to agree on and clarify the workflow with all parties at the outset of the project.

Figure 5.5 deviates slightly from BS EN ISO 19650-2; according to the standard, and given that this preliminary information was initially intended for review and comment under Status S3, the information author is required to revise it and assign a new status, Status S4. This updated information, which incorporates the comments provided in the previous S3 issue, is now ready for review and authorisation by the lead appointed party.

Following authorisation from the lead appointed party, the task team can make further revisions and issue the information to the appointing party for ultimate review and acceptance, marking it S5.

When the appointing party is satisfied and accepts the information as complete and suitable for publication, the task team will issue new information with a contractual revision and status. The information, issued with Status A5 (authorised and accepted at Stage 5), will follow the same workflow, but the process should be expedited, as the aim is to ensure that the relevant comments have been incorporated.

As this process can be quite lengthy, it can be streamlined by utilising the appropriate workflows within the common data environment (CDE) solution, and considering the reduction of the number of resubmissions, specifically for S4 and S5. In any case, ensure that the agreed information workflow is clearly communicated and approved at the start of the project with both the appointing party and the appointed task teams.

Naming protocol

In my experience, adhering to the agreed naming protocol greatly simplifies the process of locating information, reducing the need for excessive subfolders where data might get misplaced or be unable to be located. That said, for those unfamiliar with the system, the protocol can initially appear intimidating. Both the ISO 19650-2 National Annex (BSI, 2021), which includes the corrigendum from February 2021, and the UK BIM Framework (2022a, 2022b) offer comprehensive guidance on this process, complete with handy reference examples.

The importance of consistency cannot be overstated. When every document follows a well-defined naming protocol, the risk of essential data being overlooked or misplaced is reduced. Incorporating unique codes in file names can make a tremendous difference. It means that when someone is looking for specific information, they can retrieve it rapidly, rather than getting lost in a labyrinth of subfolders.

I've noted a significant evolution within the industry over recent years. Emphasis on established naming conventions has grown. When I began my journey with BIM, strict adherence to the BS 1192:2007 (BSI, 2008) naming conventions was somewhat of a rarity. However, with BIM's increased prominence, there's been a marked move towards this standard. Now the majority of the task teams supply information in line with these conventions.

There are, of course, occasional hiccups with certain task teams, but, on the whole, the industry has navigated past the initial reluctance towards these naming conventions. This acknowledgement of the importance of consistency and standardisation in information management has undoubtedly enhanced the efficacy of information management. This momentum will undoubtedly shape the future trajectory of information management within the construction and design industries.

The field codes, shown in Figure 5.6 and based on BS EN ISO 19650-2, should only use alphanumeric characters, comprising letters A-Z and digits 0-9. The standard defines each field as follows.

- *Project.* Identifies the project. The code is defined by the appointing party.
- *Originator.* The party responsible for the information. Avoid duplicated codes.
- *Functional breakdown.* Specific project function of the information, such as a system, work package, or agreed approach. Previously, it was known as Volume/System.

Figure 5.6 Field codes present in the file information container ID (Source: BSI, 2021)

- *Spatial breakdown.* Specific spatial aspect of the information, such as floor level, location, or agreed approach. Previously, it was known as Level/Location.
- *Form.* Defines the form of information, e.g., drawings (D). Previously, it was known as Type.
- *Discipline.* The technical branch that is responsible. Previously, it was known as Role.
- *Number.* Number to ensure a unique ID when considering all other fields.

A range of standard codes is specified in BS EN ISO 19650-2 and guidance is provided on how these should be used. Although this list can be expanded, it's important to ensure that any additions are clearly communicated and documented within the project's naming protocol, as the field codes should be used consistently by all of the project team.

The metadata that are recorded against each information container are critical to ensuring that all relevant information is captured and easily accessible throughout the project lifecycle, and enable us to keep track of changes and updates, as well as to categorise information for better organisation and searchability.

Two essential elements of the metadata are the revision code and the status code. The former allows stakeholders to trace the historical changes made to a document, ensuring that everyone is on the same page. The latter offers insights into where the document stands, in terms of the use for which the information container is suitable.

One area where I recommend caution is in dealing with potential confusion arising from status metadata. For example, you may come across statuses like S1, S2 and A1. It's important not to mix up these statuses – which indicate the purpose and suitability of the information container – with the statuses received after the workflow is complete, such as 'A' for accepted or 'C' for rejected.

In the UK, classification is another key metadata element for information containers within a CDE. This classification should comply with the Uniclass 2015 standards (NBS, 2023) and should make use of the project management (PM) table whenever possible, as shown in Table 5.3.

Table 5.4 lists the status codes, which indicate the suitability of the information containers within a common data environment. These codes comply with BS EN ISO 19650-2 (BSI, 2021).

Figure 5.7 shows an example to clarify how the revision and status codes change during the project when information containers are uploaded to the common data environment. Make sure that your approach is mutually agreed and understood by all project team members at the outset of the project.

Table 5.3 Common data environment metadata

Name	Description	Status	Revision	Classification
ABCD-AMA-XX-08-D-A-2300	General arrangement	A5	C04	PM_40_40 design drawings
ABCD-AMA-XX-08-D-A-5203	Ceiling details	S3	P02	PM_40_40 design drawings

Metadata according to the requirements of BS EN ISO 19650-2 (BSI, 2021)

Table 5.4 Status and revision numbers (adapted from BSI, 2021)

Work in progress (WIP)

Code	S0
Revision	P01.01, P01.02, etc. and P02.01, P02.02, etc.
Description	Information container being generated within a task team

Shared (non-contractual)

Code	S1 to S5
Revision	P01, P02, etc.
Description	S1: Information suitable for both geometric and non-geometric coordination within a delivery team
	S2: Information suitable for information or reference within a delivery team
	S3: Information suitable for review and comment within a delivery team
	S4: Information suitable for review and authorisation by a lead appointed party
	S5: Information suitable for review and acceptance by an appointing party

Published (contractual)

Code	A1, An, etc.
Revision	C01, C02, etc.
Description	Information container that has received no comments from the invited party and is either
	■ authorised by the lead appointed party
	■ accepted by the appointing party

Note: Status code 'B' (B1, Bn), which utilised Revisions codes P01, P02, etc. and was considered contractual information, is now deprecated. It referred to an information container that had been partially signed off and included comments from the invited party for either authorisation or acceptance. The 'n' is used to refer to the work stages, unless a different approach is specified in the project's information standard. Source: BSI (2021).

5.8. Measuring the benefits of BIM

Innovate UK assigned PwC the task of creating a Benefits Measurement Methodology (BMM). This was designed to measure the prospective benefits that could arise from implementing BIM Level 2 in public sector infrastructures and capital assets. Furthermore, PwC was also tasked with drafting a subsequent benefits report (PwC, 2018), detailing the anticipated benefits derived from the application of this BMM.

The benefits report elucidates the process and results derived from employing the BMM to estimate the benefits of applying BIM Level 2 to two public sector capital assets. The report tests whether the methodology is capable of estimating economic benefits that could be realised throughout the asset lifecycle.

One of the projects chosen to employ this methodology was the refurbishment of the Department of Health (DoH) headquarters, located at 39 Victoria Street, London. This particular project involved the Category B fit-out of an 11-storey building. I had the privilege of being involved in this project, leading the BIM implementation. The project took place from 2016 to 2017 and made

Figure 5.7 Revision process (see also BSI, 2021)

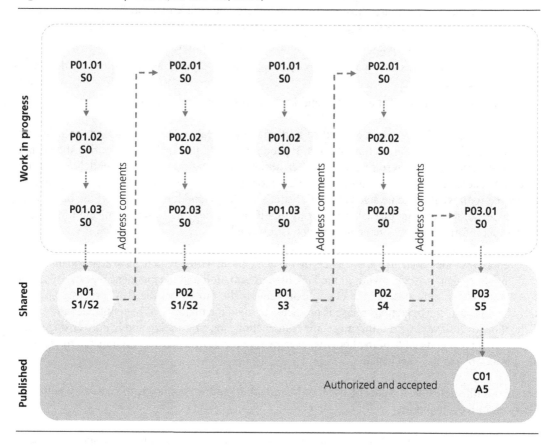

extensive use of BIM Level 2 processes, incorporating BIM during the design phase and throughout the building's operational lifespan.

This project provided us with valuable lessons and insights, particularly as it represented one of our company's most comprehensive project to date, with the traditional roles being tasked with the BIM implementation. At that time, limited examples of best practices and case studies were available in the industry for a comprehensive implementation of the BIM process from design to handover. Despite the fact that most of the team members were new to BIM, we discovered that the overall process was more manageable than we had initially anticipated. This experience allowed us to gather significant knowledge and establish a foundation for future BIM projects.

The collaboration between the main contractor, the client, their information management and project management team, and the facility management subcontractor, was crucial in delivering a BIM project that provided benefits to the client at various stages of the project. This project was one of the most successful projects I can recall, as we were able to complete it on time and within budget, leading to repeat business for the same client at other locations.

Insights of the PwC report

On reviewing the benefits outlined in the PwC report (PwC, 2018) for the DoH 39 Victoria Street project using the Benefits Measurement Methodology (BMM), several interesting points come to light.

The findings indicate significant present value total lifecycle savings of £676 907. Some may consider that these savings are substantial, while others may think that the figure is lower than expected. However, I believe this figure is important when considering the following.

- The total cost of executing the works amounted to £12.2 million.
- The operations phase of this particular project spans only 12 years, as determined by the lease agreement.
- The annual operations cost under the maintenance contract is set at £860 000.

These figures provide important context to understand the scale of the savings achieved in this project through the implementation of BIM. It should be noted that the savings during the design and construction phases, resulting from better design coordination using the models, were not taken into account. Furthermore, the structure used for the data in this project was relatively simple compared with what could be achieved with better knowledge and tools to support the operations phase. Hence, it is evident that BIM can bring significant savings to all phases of the lifecycle if the correct information management approach is followed.

The report provides conclusive evidence of the benefits of BIM, which have historically been challenging to demonstrate, owing to the lack of substantiation and independent reports, by using a real example. It indicates that the client team benefits the most from the BIM process. Specifically, in this case, the operations phase is predicted to yield the majority of benefits, accounting for 73% of the overall savings. Furthermore, the construction, commissioning and handover stages are expected to contribute 21% of the benefits. Meanwhile, the design phase is anticipated to contribute a minor 6% to the total benefits.

Figure 5.8 illustrates the benefits of BIM, as identified in the PwC report, categorised by both category and lifecycle phase.

The distribution of these benefits isn't surprising, as past research has shown that operational and maintenance expenses can make up 60% to 80% of the total costs over a project's lifespan, which typically lasts around 30 years. Therefore, projects with a longer lifespan will reap more benefits during the operation and maintenance phase from the use of BIM. Although the 39 Victoria Street project has a shorter lifespan, we can still observe the advantages that the operations and maintenance team can gain from the use of BIM.

In past training sessions I've conducted, it has become apparent that many people in the construction sector are not fully aware of the costs tied to operations and maintenance throughout a project's life. The initial focus in our sector, from a main contractor's point of view, is predominantly on the design and construction phases, which can obscure the long-term operational and maintenance expenses that clients will face. It is crucial to raise awareness about these frequently overlooked costs. By doing so, we can achieve a comprehensive understanding of the project's complete financial footprint. Moreover, this highlights the importance of providing accurate information to the end user once a project is completed.

The Royal Academy of Engineering paper by Evans *et al.* (1998) suggests that, for an office building over 30 years, the cost ratio of initial construction to maintenance and operation is 1:5. While there may be debates over the exact figures, this can be used as a rule of thumb, as the ratio can vary as a result of many factors unique to the project. However, the overriding message is

Figure 5.8 Benefits by category and lifecycle phase (data from PwC, 2018)

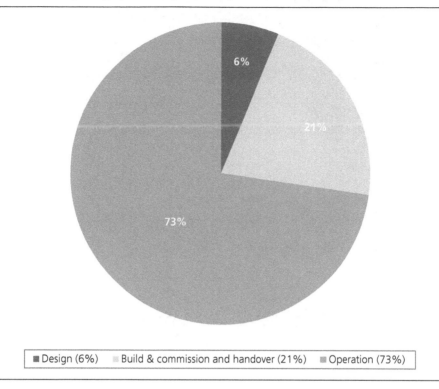

■ Design (6%)   ■ Build & commission and handover (21%)   ■ Operation (73%)

clear: operations and maintenance account for the most significant part of the total lifecycle cost. Figure 5.9 shows that, for a 30 year project lifecycle, design and construction represent 20–40% of the cost, while operations and maintenance comprise 60–80% of total costs, as referred to by Rounds (2018).

A well-developed strategic operations and maintenance plan at the start of the project, with the correct use of accurate data collected during the different phases of the project, is crucial to support clients and end users during the building's operational phase, and it is here where we need to increase focus to unlock larger benefits.

Therefore, it is not surprising to see that the largest benefits of BIM, as estimated within the PwC (2018) report, are at the handover stage and during the operations and maintenance of the building. It is worth noting that the benefits measured are based on a shorter lifespan; however, we can still observe the advantages that the operations and maintenance team can gain from the use of BIM.

Consideration of clash detection benefits

The advantages and cost reductions associated with clash detection were not considered in the creation of this report. When we prepared our contribution to the report, we decided not to estimate these costs, to avoid the possibility of skewing the final results. At that time, there was no established method for accurately quantifying these benefits and savings. However, we believe that

Figure 5.9 Lifecycle cost (data from Rounds, 2018)

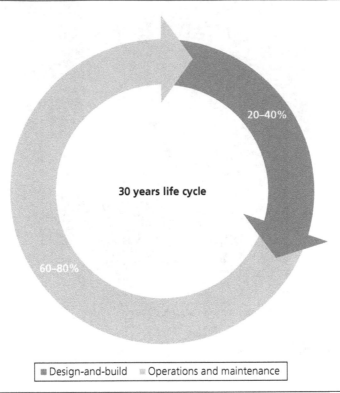

20–40%

30 years life cycle

60–80%

■ Design-and-build   ▫ Operations and maintenance

including the effects of clash detection during the design and construction stages of this project would significantly increase the final cost savings of implementing BIM.

Many of the design mistakes we identified during the initial use of the BIM process could also have been detected through more conventional methods. The traditional process might take longer, however, and not all issues might come to light. However, it is particularly challenging to determine where the exclusive benefits of BIM begin.

While I was convinced that many of the clashes identified and resolved during the design and construction phase using the 3D model would not have been identified using traditional methods, I had no evidence to support this claim. Therefore, we decided to exclude these scenarios from the report. Additionally, I believe that it is quite challenging to measure the cost saved through clash detection or design coordination in a 3D design, owing to the multitude of variables involved.

Savings in operations and maintenance

The use of BIM in the design and construction phases has undoubtedly granted substantial advantages. However, I would like to emphasise the significance of the benefits derived from the operations and maintenance phase, highlighted in the PwC report. The significance of benefits at this stage is often overlooked in the construction industry. The feedback from the facility management team featured in this report offers valuable insights from end users on how BIM implementation would support them on this project.

As detailed in the report, the facility management subcontractor was contracted just before the conclusion of the construction phase. They revised the asset register and aligned the COBie data with the entries in the computer-aided facility management (CAFM) system. Although, as I have covered in the previous chapter, this is not the best approach to maximise the benefits of the BIM process. Ideally, they should have been involved from the design stage. Nevertheless, the information provided helped reduce the need for the additional asset validation surveys usually conducted by the facility management to assess the actual condition of assets.

The Department of Health (DoH) committed to maintaining the asset information model (AIM) after project completion and undergoing necessary updates. This model enables effective building management and informs future alterations to the space layout or maintenance operations.

Although Appendix B in the PwC report offers a thorough assessment of the estimated savings benefits during the different phases of the project, including the design-and-build and commission stages, my interest lies in exploring the benefits during the handover and operational stages to provide more visibility of these benefits.

- Time savings in asset handover processes (£84 520)
  - *Efficient asset validation surveys impact.* Using an accurate asset information model eliminates the need for regular, full surveys by facility management (FM) contractors. Without this model, staff site visits to validate assets would occur roughly every 5 years, based on the contract with the Department of Health.
  - *Swift import to CAFM systems impact.* The asset information model allows for a faster handover of up-to-date data than can be achieved by updating 2D diagrams. Inputting COBie data throughout the design phases also speeds up the integration of asset data in the CAFM system, compared with manual entries.
- Cost savings in asset maintenance (£391 592)
  - *Reactive maintenance cost savings with BIM.* Using BIM provides clearer information for reactive maintenance, increasing the likelihood of solving issues on the first attempt. Without BIM, additional attempts can raise costs. Stakeholders suggest potential savings of about 10% annually.
  - *Regular maintenance efficiency through BIM.* Better clash detection with BIM can reduce regular maintenance challenges, such as the accessibility issues often found with fire dampers. This reduces the need for additional equipment or adjustments, considering the yearly inspections required for numerous dampers.
  - *Optimised maintenance and extended asset life with BIM.* Using an asset information model optimises maintenance over an asset's lifespan. Without BIM, about 20–30% of non-statutory planned maintenance is often overlooked, reducing asset life by up to 20%. With BIM, while annual non-statutory maintenance costs might rise, the total expenditure (TOTEX), namely capital expenditure (CAPEX) plus operational expenditure (OPEX), is reduced, because of the less frequent requirement to replace asset components, leading to long-term savings.
- Cost savings in refurbishment (£23 463)
  - Using an asset information model during refurbishments streamlines postcommissioning design changes. A detailed 3D model reduces design team involvement and ensures pre-emptive clash detection before on-site work.

- Decreased variation in operating expenditure (£2943)
  - ☐ The handover documentation's AIM and COBie data enhanced the FM contractor's trust in the asset register's accuracy. This improved precision in predicting O&M costs and reduced associated risk by an estimated 10%.
- Improved asset utilisation (£28151)
  - ☐ Using BIM data reduces the likelihood of unexpected building shutdowns by providing a detailed asset register to guide FM activities. This clarity on the condition of vital equipment minimises disruptions. Stakeholders have highlighted the benefits of BIM in such scenarios.

Figure 5.10 breaks down the value of benefits at different lifecycle phases, as detailed in the results from the PwC report.

Reinforcing the focus on supporting the operational phase

Throughout history, the construction industry has predominantly focused on the design and construction stages of projects. However, there has been a significant oversight in terms of the quality of the information that we pass on to our clients. We have often prioritised short-term cost savings and value engineering, neglecting to fully consider the long-term implications and sometimes deviating from the original specifications. It is crucial for project teams to fully understand these implications and provide accurate and precise information.

Figure 5.10 Value of benefits by lifecycle phase (adapted from PwC, 2018)

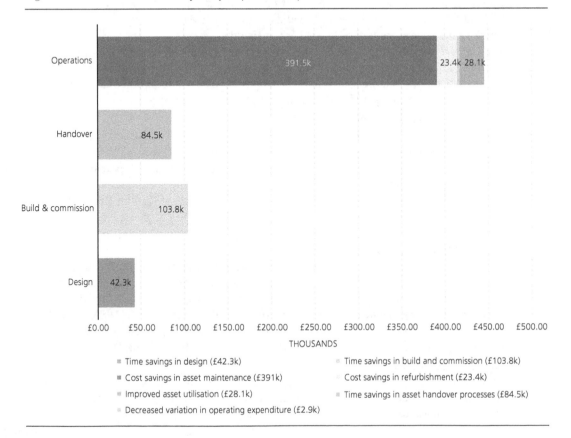

Unfortunately, there is a lack of comprehensive and honest case studies from end users, sharing their perspectives on how the implementation of BIM has aided them during the operational phase. To enhance our understanding of the benefits of BIM and identify areas for improvement to increase efficiency and deliver higher quality information in future projects, we require more feedback from end users. The PwC (2018) report and its findings, therefore, present an excellent opportunity and tool for industry professionals. The report provides tangible evidence of the advantages that BIM can bring to clients. If clients can grasp these benefits and insist on the implementation of BIM in their projects, this will accelerate the ongoing transformation of the industry.

The information presented in the report should encourage clients to demand superior standards and hold stakeholders accountable for the quality of the information provided. The recently introduced Building Safety Act 2022 (HMG, 2022), which includes specific data requirements, can only be effectively implemented if these requirements are seamlessly integrated in a robust and well-practised business process that revolves around high-quality information management.

Building on the earlier discussion about the importance of data quality and information management, it's clear that in recent years, technology has primarily been leveraged to assist in the design and construction phases, as these stages are more tangible and easily understandable to all stakeholders involved. While the significance of technology in design and construction decisions should not be underestimated, there needs to be a shift in focus towards technology that enables the automation and digitalisation of processes. This will facilitate the capture and delivery of accurate, verified and validated information to clients, assisting stakeholders during the operations and maintenance phase and effectively managing information and data. By doing so, we can maximise the associated benefits while reducing reliance on human input, which often leads to errors and inconsistencies.

To ensure simplicity and clarity in asset information management, it is crucial to address three essential questions.

- What is the asset?
- Why is it important to capture data for the asset?
- What information is needed to maintain the asset? For example, location, serial number, installation date, supplier, manufacturer details, SFG20 codes or specific attributes.

Translating these questions into practical outcomes that align with the needs of end users is of utmost importance, as it ensures that the data capture is both necessary and done correctly. Data capture is meaningless unless it is meaningfully applied within the broader context of end user operations. Embracing simplicity is even more critical, as some clients may strive to transition rapidly from a minimal data environment to an overwhelming one, without having an appropriate strategy in place. This can result in requests for irrelevant information that does not serve their specific assets.

In addition to the benefits reaped by clients, a focus on accurate information management during projects will have a positive and far-reaching impact on the industry as a whole. When clients prioritise the delivery of accurate and reliable information to efficiently manage their assets, this serves as an impetus for businesses to take information management seriously. This, in turn, helps safeguard the industry, enhances our collective reputation and cultivates trust among stakeholders.

5.9. Collaborate instead of judging

The digital construction industry is constantly evolving and disagreements and misunderstandings are inevitable. However, it is essential to approach them collaboratively rather than judgementally. Even those who market themselves as industry leaders can make mistakes. In fact, when we delve deeper, we often find that those who claim to be experts and the most knowledgeable have gaps in their collaboration with stakeholders and the execution of the project.

Therefore, imposing your own principles and processes and acting as if you have the only truth is not an effective way to work together. Humility and collaboration are key to achieving success in this industry – we are all here on the same boat. We are all still learning how to improve the delivery of the BIM process, especially considering the varying levels of adoption among stakeholders and different types of projects.

Implementing the BIM process may seem straightforward, but it's far from easy, particularly when critical stakeholders hold a negative outlook towards change. It's essential to avoid overcomplicating the documentation or duplicating information and to focus on keeping the content clear and correct. Otherwise, it's easy for things to become overly contractual and for relationships to become strained when challenges arise, especially when the skills and resources promised during the tender stage don't fully materialise.

Teams need to have the correct information to support the project and protect the business. It's also crucial to understand what you are committing to do and the level of information you will be held accountable for developing and delivering throughout the project. We don't get paid by the number of words in the BIM documentation but by having concise and correct information to support the project and protect the business in the most effective way.

There are times when task teams at the tender stage exaggerate the challenges associated with the quality of information or level of coordination received from other teams or even the time and effort needed to name files according to the naming protocol to support the correct management of information in the common data environment. It's important to keep in mind that we're all working towards the same goal: delivering a successful project.

It's important to remember that while some complaints may be legitimate, there are also instances where the sole aim is to gain more time and increase fees by complaining about the quality of information received. To avoid these situations, it's crucial for companies to thoroughly review the information they receive and ensure that the quality meets the project standards before passing it on to other task teams. Some may try to take advantage of others' lack of knowledge or understanding to gain a commercial advantage, and it's important to be aware of this possibility. Having the knowledge and skills to understand the validity of their complaints is key to addressing these difficult conversations, offering advice and support. Remember, we're all working towards the same goal: delivering a successful project.

It's essential for companies to understand their own capabilities and limitations to avoid taking on more than they can handle. We should strive for transparency and honesty in our business dealings, and work towards building trust and respect with our colleagues and partners.

Implementing digital construction in a company and working with external stakeholders can be challenging and exhausting, but it's essential to keep an open mind and observe what others are saying and doing. Be transparent and honest about your experiences, don't hesitate to seek help when needed, and share the lessons you've learned with others. Understanding the benefits of the BIM process during a project, as well as its long-term impact on facility management during the operation and maintenance phase, is crucial for removing barriers and convincing those who are sceptical about its value. Ultimately, by working together collaboratively and with a supportive mindset openly, we can deliver successful projects and improve the industry as a whole.

BIBLIOGRAPHY

BSI (2016) BS 1192:2007+A2:2016: Collaborative production of architectural, engineering and construction information. Code of practice. BSI, London, UK.

BSI (2013) PAS 1192-2:2013: Specification for information management for the capital/delivery phase of construction projects using building information modelling. BSI, London, UK.

BSI (2019) BS EN ISO 19650-1:2018: Organization and digitization of information about buildings and civil engineering works, including building information modelling (BIM). Information management using building information modelling. Part 1: Concepts and principles. BSI, London, UK.

BSI (2020) BS EN ISO 19650-5:2020: Organization and digitization of information about buildings and civil engineering works, including building information modelling (BIM). Information management using building information modelling. Part 5: Security-minded approach to information management. BSI, London, UK.

BSI (2021) BS EN ISO 19650-2:2018 & Revised NA: Organization and digitization of information about buildings and civil engineering works, including building information modelling (BIM). Information management using building information modelling. Part 2: Delivery phase of the assets. BSI, London, UK.

ISO (2022) ISO/IEC 27001:2022: Information technology. Security techniques. Information security management systems. Requirements. BSI, London, UK.

Cabinet Office (2011) *Government Construction Strategy.* Cabinet Office, London, UK. https://assets.publishing.service.gov.uk/media/5a78ce8eed915d07d35b2933/Government-Construction-Strategy_0.pdf (accessed 23/11/2023).

Evans R, Haryott R, Haste N and Jones A (1998) *The Long Term Costs of Owing and Using Buildings.* Royal Academy of Engineering, London, UK.

HMG (Her Majesty's Government) (2022) Building Safety Act 2022. The Stationery Office, London, UK.

IPA (Infrastructure and Projects Authority) (2021) Transforming Infrastructure Performance: Roadmap to 2030. https://www.gov.uk/government/publications/transforming-infrastructure-performance-roadmap-to-2030/transforming-infrastructure-performance-roadmap-to-2030 (accessed 23/11/2023).

National Cyber Security Centre (2023) About Cyber Essentials. https://www.ncsc.gov.uk/cyberessentials/overview (accessed 16/11/2023).

National Protective Security Authority (2023) Security-Minded Approach to Information Management. https://www.npsa.gov.uk/security-minded-approach-information-management (accessed 24/11/2023).

NBS (2023) Uniclass. https://uniclass.thenbs.com/ (accessed 27/11/2023).

PwC (2018) *BIM Level 2 Benefits Measurement: Application of PwC's BIM Level 2 Benefits Measurement Methodology to Public Sector Capital Assets.* PwC, London, UK. https://www.cdbb.cam.ac.uk/files/4.pwcbmmapplicationreport_0.pdf (accessed 23/11/2023).

Rounds D (2018) Design for maintainability: the importance of operations and maintenance considerations during the design phase of construction projects. *Whole Building Design Guide.*

https://wbdg.org/resources/design-for-maintainability#:~:text=Studies%20show%20that%20 operation%20and,maintenance%20tasks%20within%20that%20system (accessed 23/11/2023).

SFG20 (2023) What is the SFG20 Standard? https://www.sfg20.co.uk/what-is-sfg20 (accessed 16/11/2023).

UK BIM Framework (2022a) ISO 19650 Guidance C: Facilitating the CDE (Workflow and Technical Solutions). https://ukbimframeworkguidance.notion.site/ISO-19650-Guidance-C-Facilitating-the-CDE-workflow-and-technical-solutions-ff3bdbcf1c1349c1a98c586943d0 a9f1 (accessed 24/11/2023).

UK BIM Framework (2022b) ISO 19650 Guidance F: Information Delivery Planning. https:// ukbimframeworkguidance.notion.site/ISO-19650-Guidance-F-Information-delivery-planning-dbc32ce51f2c4c598d6ae24af06adda2 (accessed 24/11/2023).

Amador Caballero
ISBN 978-1-83549-446-2
https://doi.org/10.1680/iceedc.9446206

Chapter 6
Strategy, roles and procurement

6.1.   Introduction

In this chapter, I explore the complexities of managing information during the asset delivery phase, in accordance with ISO 19650-2 (BSI, 2021). This particular standard provides a thorough set of guidelines for information management and outlines the necessary steps for effective collaboration between the appointing party, the lead appointed party and various task teams. This ensures successful information management throughout the project.

The chapter begins by detailing the duties assigned to each role in the delivery team, such as the lead appointed party and the task teams, as outlined in Section 2.9. The aim is to clarify how the requirements of ISO 19650-2 are implemented in these roles. Understanding this is essential for effective information management across the entire construction process, benefiting both the lead appointed party and the involved task teams.

The chapter features two sections specifically focused on the appointing party, offering a thorough examination of the particular responsibilities of this role.

First, it stresses the need for appointing parties to establish an information management strategy within their organisations. This strategy should also define the information management function, which plays a vital role in effectively managing information throughout its lifecycle.

Second, the chapter reviews the relevant clauses in ISO 19650-2 that concern the appointing party, ranging from the assessment and needs phase to the project close. The aim is to break down the key clauses into easily digestible guidance for those keen on understanding the responsibilities of the appointing party according to ISO 19650-2.

Similarly, in this chapter I explain the specific clauses related to the prospective lead appointed party during the tender response and the appointment phase. These details are laid out in a user-friendly manner for easy reference. The section specifies the kinds of document required at each stage and explores their contents to guide you in generating your own documents. A parallel approach is applied to the tendering and appointment of various task teams in the following section. The goal of both sections is to assist you in navigating the complexities of these clauses and offer valuable insights in preparing the necessary documentation.

The chapter concludes by comparing traditional procurement methods with the design-and-build approach. It outlines considerations for these types of projects during the tendering phase, thereby giving a comprehensive overview.

6.2.   Process and impact on roles

In this section, I describe the actions relevant to the lead appointed party and the task teams during the information management process, as outlined in ISO 19650-2 (BSI, 2021).

I have mapped the various roles and responsibilities within the lead appointed party for each clause. This is to assist in understanding the new responsibilities of the roles engaged in the information management process, including delivery on site.

It's important to note that this information is intended solely as a reference. Specific responsibilities and tasks may vary, depending on the nature of the business, internal processes, size and procurement methods used.

This guidance does not replace the need to read ISO 19650-2 to fully understand the details of each clause.

Tender response (Clause 5.3)

A summary checklist of the tender response team actions is given in Table 6.1.

Table 6.1 Checklist: tender response team actions (continued on next page)

Nominate individuals to undertake the information management function (Clause 5.3.1)
Preconstruction manager
- ☑ Take charge of the tender information review.
- ☑ Make sure that digital construction lead is invited to tender launch and understands project details.

Digital construction lead
- ☑ Attend the tender launch.
- ☑ Review and assess tender information for opportunities and risks.
- ☑ Share findings with preconstruction team.

Establish the delivery team's (pre-appointment) BIM Execution Plan (Clause 5.3.2)
Digital construction lead
- ☑ Clarify project requirements by reviewing customer information and raising tender queries.
- ☑ Produce the pre-appointment BIM Execution Plan, based on customer's exchange information requirements and task team input, and include it as part of the tender response.
- ☑ Take initiative to produce BIM Execution Plan if customer has not provided any BIM requirements.

Assess task team capability and capacity (Clause 5.3.3)
Estimators and commercial team
- ☑ Consider the BIM capabilities of task teams during tender process.
- ☑ Provide BIM documentation and BIM assessment questionnaire to relevant task teams.
- ☑ Provide access to models for task teams to review and quote.

Digital construction lead
- ☑ Clarify BIM requirements with task teams.
- ☑ Seek clarification from the appointing party over any concerns raised by task teams.

Task team
- ☑ Complete the BIM assessment questionnaire honestly.
- ☑ Review the BIM documentation and provide feedback to the lead appointed party.

Table 6.1 Continued

Establish the delivery team's capability and capacity (Clause 5.3.4)
Digital construction lead
- ☑ Review and assess BIM questionnaires from task teams.
- ☑ Identify gaps or training needs.
- ☑ Provide capability and capacity assessment summary as part of the tender response.

Design manager and technical services manager
- ☑ Consider BIM assessment results and recommendations from the digital construction lead.
- ☑ Evaluate previous performance scores and notes.
- ☑ Make a selection based on capabilities, not solely on cost.

Establish the delivery team's mobilisation plan (Clause 5.3.5)
Digital construction lead
- ☑ Collaborate to create a mobilisation plan as part of the tender response.
- ☑ Agree on information production methods and procedures.
- ☑ Conduct necessary tests for software interoperability.

Establish the delivery team's risk register (Clause 5.3.6)
Digital construction lead
- ☑ Identify and communicate BIM-related risks as part of the tender response.

Design manager
- ☑ Incorporate risks and comments in the project risk register.
- ☑ Incorporate feedback from other task team members in the risk register.
- ☑ Include lessons learned from previous projects.

Compile the delivery team's tender response (Clause 5.3.7)
Preconstruction manager
- ☑ Consult the digital construction lead for the ITT documentation.

Digital construction lead
- ☑ Answer any questions related to BIM and provide business case as part of the tender response, if required.

Nominate individuals to undertake information management (Clause 5.3.1)
The prospective lead appointed party will nominate individuals within the organisation or appoint another party to manage information on its behalf. If a third party is appointed, the lead appointed party will define their scope of services.

PRECONSTRUCTION MANAGER
Once the tender information from the appointing party becomes available, it's essential for the preconstruction manager to ensure that the digital construction lead is fully briefed on the project details. The digital construction lead should also be invited to the tender launch discussion.

This invitation provides an excellent opportunity for the digital construction lead to acquire an in-depth understanding of the project's overall details and requirements and to meet the delivery team and collaborate with other stakeholders involved in the project. Through such collaboration, the team can identify any potential issues that may crop up during the project and find ways to mitigate them during the tender stage.

If the appointing party has not yet considered implementing BIM, and the project is still in its early conceptual and design stages, then it falls on the preconstruction manager and the digital construction lead to advocate for its adoption. They should articulate the benefits of adopting BIM, highlighting how it can offer advantages at various stages of a project's lifecycle and to the organisation as a whole. Support and guidance should be extended to the appointing party to facilitate the implementation of BIM.

DIGITAL CONSTRUCTION LEAD

To ensure effective information management for a project, it's important for the digital construction lead to be involved from the earliest stages. Therefore, attendance at the tender launch is essential for the digital construction lead to acquire the correct understanding of the project and offer appropriate support on any new opportunities.

As soon as the tender information becomes available, the digital construction lead should carefully review and assess the provided details. This involves identifying any opportunities or risks related to digital construction delivery for the project. All findings, concerns and thoughts should be promptly communicated to the preconstruction team.

If no BIM documentation has been provided during the tender stage, the digital construction lead should take the initiative to generate the necessary information. This will assist the appointing party in adopting the BIM process or in complying with business policy. The current stage of the project should be considered while generating this information.

In the context of a traditional contract, the digital construction lead should evaluate the quality of the project deliverables provided by the design team. This evaluation should be compared against the BIM Execution Plan, if one exists, and should consider the process checks outlined in Section 6.7.

Establish the delivery team's (pre-appointment) BIM Execution Plan (Clause 5.3.2)

The prospective lead appointed party will produce and include the delivery team's pre-appointment BIM Execution Plan in its tender response.

DIGITAL CONSTRUCTION LEAD

During the tender stage, the digital construction lead needs to ensure that all project requirements are clearly defined, to avoid any assumptions. To accomplish this, the digital construction lead will raise any necessary tender queries to clarify ambiguous or unclear project requirements. This step is essential for ensuring that the delivery team has a comprehensive understanding of the appointing party's objectives and requirements, thereby helping to avert misunderstandings and errors later in the project lifecycle.

In addition, the digital construction lead is tasked with creating the appropriate pre-appointment BIM Execution Plan. This plan is based on the exchange information requirements and other available documentation provided by the appointing party and is developed with input from the known task teams. This method ensures that the delivery team clearly understand their responsibilities to meet the appointing party's expectations.

If the appointing party has not provided any exchange information requirements or has not set any BIM expectations for the project, the digital construction lead will proactively create the BIM Execution

Plan. This will be done in accordance with both business policy and industry best practices to facilitate effective project delivery and to compile the relevant information needed for future business operations.

Assess task team capability and capacity (Clause 5.3.3)

The task teams involved in the tender process will be evaluated based on their capability and capacity to meet the exchange information requirements and the proposed pre-appointment BIM Execution Plan.

ESTIMATORS AND COMMERCIAL TEAM

During the tender process, both the estimators and the commercial team should take into account the BIM capabilities of different task teams, including design consultants and subcontractors. This consideration aids the preconstruction team in selecting those most apt for successful project delivery.

The estimator and commercial team should provide BIM documentation, including any available models and other relevant information, to all pertinent parties during the tender process. This will help clarify BIM expectations for the project.

The BIM capability and capacity assessment for managing and delivering information must be completed as part of the task team's tender response.

Note that BIM documentation and assessments should target companies based on their specific responsibilities. This streamlines the tender process by avoiding unnecessary assessments and the provision of excessive information.

DIGITAL CONSTRUCTION LEAD

The digital construction lead needs to make certain that BIM requirements are clearly communicated to the task teams, address any questions or concerns and, if needed, seek further clarification from the appointing party during the tender stage. This approach helps maintain uniform understanding among all parties, reducing potential misunderstandings or disagreements later in the project. It also secures accurate cost estimates from each task team, negating the need for additional costs to cover unforeseen risks.

TASK TEAM

Task teams are each responsible for completing the BIM assessment honestly and to the best of their ability. This provides the lead appointed party with a transparent understanding of each task team's capabilities in terms of BIM requirements. It ensures that task teams are assigned roles they can effectively execute, thus minimising the risk of delays and additional costs. Additionally, it allows the lead appointed party to offer any necessary training to support the task team when needed.

In addition to completing the BIM assessment questionnaire, task teams should meticulously review the provided BIM documentation. They should then offer any feedback or concerns to the lead appointed party. This ensures that any issues can be raised and resolved in a timely manner, confirming that project requirements are clearly understood and potential obstacles are addressed before they escalate into major problems.

Establish the delivery team's capability and capacity (Clause 5.3.4)

The prospective lead appointed party will assess the task teams' responses and compile a summary of their skills. This summary will also identify any extra training needed to effectively manage and produce information as required. This assessment summary will be included in the tender response submitted to the appointing party.

DIGITAL CONSTRUCTION LEAD

Once the BIM assessment questionnaires have been completed by the different task teams and provided to the lead appointed party as part of the tender response, it is the responsibility of the digital construction lead to carefully review and assess the answers and evidence provided. This step will identify any gaps or training needs within the task team capabilities and make recommendations to the preconstruction team regarding the selection of appropriate task teams for the project.

It's worth noting that, as part of the tender response, a capability and capacity assessment summary needs to be provided to the client. This requirement provides the appointing party with awareness of the capabilities and capacities of each task team to deliver on the BIM requirements for the project.

DESIGN MANAGER AND TECHNICAL SERVICES MANAGER

As part of the task team selection process, it is crucial for design managers and technical services managers to take into account the BIM assessment results and recommendations provided by the digital construction lead. These assessments offer valuable insight into the capabilities of each task team and their ability to meet the project's BIM requirements.

However, while the assessment scores based on the responses provided at the tender stage are useful, the preconstruction team also needs to consider the performance of the task teams in previous similar projects. This can help identify weaknesses or strengths of the task teams, enabling better decision making in the selection process.

It's important to note that the selection of the task team should not be based solely on cost. Instead, it should focus on the team's ability to deliver the project efficiently while meeting BIM requirements. Therefore, the preconstruction team must stand firm during settlement discussions to ensure that the most appropriate selection is made.

By carefully considering all relevant information, including BIM assessments and past performance, design and technical services managers can make the best recommendation for selecting a task team with the necessary capabilities and expertise to efficiently and effectively deliver on the project requirements.

Establish the delivery team's mobilisation plan (Clause 5.3.5)

The prospective lead appointed party will create a mobilisation plan for the delivery team. This mobilisation plan will be included in the tender response submitted to the appointing party.

DIGITAL CONSTRUCTION LEAD

The digital construction lead, in collaboration with the rest of the delivery team, will create a mobilisation plan. This plan includes determination of various methods and procedures for information production, setting a naming convention, providing specific training, establishing the correct survey strategy and specifying the digital tools to be implemented for the project.

To ensure proper information exchange and software interoperability, the project team will conduct the necessary tests. These tests confirm that the agreed digital tools are functional and that the project team can use them effectively. Successful implementation of the mobilisation plan helps to guarantee that the project team will meet the project's needs efficiently.

Establish the delivery team's risk register (Clause 5.3.6)

The prospective lead appointed party will produce a risk register, identifying risks associated with the delivery of information on time, and outline how the team intends to manage them. These risks should correspond to the information exchange information requirements set by the appointing party. Every member of the delivery team will contribute to this risk register, which will be included in the tender response submitted to the appointing party.

DIGITAL CONSTRUCTION LEAD

The digital construction lead will identify and communicate any risks linked to the BIM process. These risks could affect the delivery of the appointing party's exchange information requirements and project milestones. For smooth and efficient project delivery, it's crucial to identify any issues concerning the information protocol and any unaddressed assumptions during the tender query process with the appointing party.

Additionally, the digital construction lead should flag any other potential obstacles to the proposed information delivery strategy. This includes considerations of point cloud survey requirements, the specific technology to be used and the level of information and asset data needed.

If some task teams are novated, or if some information has already been produced, the digital construction lead is responsible for communicating any concerns. This includes the capabilities and experience of the novated teams and any discrepancies or errors within the information produced that might fall short of the client's expectations.

Effective risk management for the BIM process also requires the digital construction lead to consider feedback from other task team members. This input should be included in the overarching project risk register. While a single, comprehensive risk register is required, I personally include BIM-related risks in both the BIM Execution Plan and the overall project risk register. This ensures easy access to vital information, but it's important to keep all documents consistent and accurate.

DESIGN MANAGER

The design manager's responsibilities include compiling the risks and comments provided by the digital construction lead and other task team members, not limited to BIM, in the overall project risk register. This consolidated information should then be shared with the client to indicate potential risks and management strategies during the tender stage.

Moreover, the design manager should use insights from lessons learned in previous projects. These can offer valuable considerations for the current project's delivery. By incorporating these lessons, the design manager aids in mitigating potential risks and contributes to the project's successful delivery.

In some companies, the responsibility to amalgamate the overall risk register may fall to the preconstruction manager.

Compile the delivery team's tender response (Clause 5.3.7)

The prospective lead appointed party is responsible for ensuring that the tender response includes the necessary documentation and covers the client's project expectations.

PRECONSTRUCTION MANAGER

As part of the invitation to tender (ITT) response, the preconstruction manager works closely with the digital construction lead, to address any specific questions about business capabilities and expertise and to fulfil the BIM expectations for the project.

DIGITAL CONSTRUCTION LEAD

The digital construction lead will supply the necessary evidence from previous projects to support the ITT, along with the required documentation to be included in the tender response. This involves providing the pre-appointment BIM Execution Plan, in accordance with the exchange information requirements, contributing to the overall risk register by providing relevant input, outlining the mobilisation plan and submitting a summary of capability and capacity assessments.

During the tender stage, the digital construction lead works closely with the design manager, operations manager and task teams. The objective is to develop the most efficient and time-effective survey strategy to meet the project's needs, while also identifying any potential issues in the available design. The digital construction lead will further assist in pinpointing any specific processes or technologies that may be necessary for successful project delivery.

Appointment (Clause 5.4)

A summary checklist of the appointment team actions is given in Table 6.2.

Confirm the delivery team's BIM Execution Plan (Clause 5.4.1)

The lead appointed party will update and confirm the BIM Execution Plan in collaboration with each task team.

DIGITAL CONSTRUCTION LEAD

The BIM launch meeting marks a milestone in the project's BIM implementation process. This meeting establishes the groundwork for information management across the project and ensures that all project team members understand the BIM Execution Plan.

The BIM launch meeting can either be a stand-alone event or part of the first design team meeting. This depends on the complexity and scope of the project.

The key representatives of the appointing party, and the task teams are required to attend. This ensures that the BIM Execution Plan is consistent with the project's design, construction and operational requirements. It also helps team members understand the actions and responsibilities needed for successful BEP implementation.

Attendance by the appointing party, as a major stakeholder with a vested interest in the project's success, is also recommended. This input is invaluable for tailoring the BIM Execution Plan to meet specific requirements and expectations.

During the BIM launch meeting, the lead appointed party will discuss any updates required to the BIM Execution Plan with the team. These updates are based on feedback received during the tender response stage. This confirms that any assumptions made or changes that have occurred since the initial, pre-appointment, BIM Execution Plan are properly addressed and incorporated. Members responsible for BIM within each task team will participate in the discussion, offering their insights. Additionally, project leads will be on hand to answer questions and clarify roles and expectations.

TASK TEAM

Each task team is responsible for attending the BIM launch meeting and rigorously reviewing the BIM Execution Plan. Teams should also provide any information the lead appointed party requires to update the plan. This ensures that the updated BIM Execution Plan effectively outlines the team's approach to information management and technology use in project delivery. The final plan

Table 6.2 Checklist: appointment team actions

Confirm the delivery team's BIM Execution Plan (Clause 5.4.1)
Digital construction lead
- ☑ Conduct BIM launch meeting with participation of key stakeholders.
- ☑ Review and update the project BIM Execution Plan with selected task teams.
- ☑ Clarify roles and responsibilities in the meeting.

Task team
- ☑ Attend the BIM launch meeting.
- ☑ Review the BIM Execution Plan.
- ☑ Provide required information for updating the BIM Execution Plan.

Establish the delivery team's detailed responsibility matrix (Clause 5.4.2)
Digital construction lead
- ☑ Create a detailed responsibility matrix.

Design manager and technical services manager
- ☑ Collaborate in the production of the detailed responsibility matrix.
- ☑ Clarify roles, milestones and dependencies in the matrix.

Establish the lead appointed party's exchange information requirements (Clause 5.4.3)
Digital construction lead
- ☑ Define exchange information requirements for each appointed party.
- ☑ Make sure requirements meet both the appointing party's needs and any additional needs.

Establish the task information delivery plan (Clause 5.4.4)
Task team
- ☑ Develop and maintain a task information delivery plan (TIDP).

Design manager and technical services manager
- ☑ Review the TIDP provided by each task team.
- ☑ Pay attention to details, such as drawing setups and naming protocols.
- ☑ Clarify the 'first issue date' and 'delivery date' for documents.

Commercial
- ☑ Incorporate the TIDP in the appointments.

Establish the master information delivery plan (MIDP) (Clause 5.4.5).
Design manager
- ☑ Manage and maintain the MIDP.
- ☑ Consolidate individual TIDPs into the MIDP.

Complete the lead appointed party's appointment documents (Clause 5.4.6)
Preconstruction manager
- ☑ Review and manage key documents to be included in the lead appointed party's appointment.
- ☑ Manage changes through change control.

Complete the appointed party's appointment documents (Clause 5.4.7)
Commercial
- ☑ Understand compliance requirements for each task team.
- ☑ Include essential documents in appointments, managing them through change control.

should offer clear processes and relevant information to satisfy the appointing party's expectations regarding information management and technology use.

Establish the delivery team's detailed responsibility matrix (Clause 5.4.2)

The lead appointed party will create a detailed responsibility matrix, specifying what information is to be delivered, when it should be distributed and which task team is responsible for creating it.

DIGITAL CONSTRUCTION LEAD

With support from the design manager and technical services manager, the digital construction lead will create the detailed responsibility matrix specific to the project.

DESIGN MANAGER AND TECHNICAL SERVICES MANAGER

The design manager, technical services manager and other team members collaborate to develop this matrix, based on their specific roles. The matrix highlights key delivery milestones, who is responsible for what, how things will be done and what factors need to be considered in the planning stages.

The detailed responsibility matrix is essential for two main reasons. First, it removes any confusion or assumptions about what information is needed. Second, it ensures that everyone knows their specific duties for delivering this information both on time and accurately. In short, the detailed responsibility matrix is crucial for the success of the project, making sure everyone is clear on their roles and accountable for their tasks.

Establish the lead appointed party's exchange information requirements (Clause 5.4.3)

The lead appointed party will determine the exchange information requirements for each task team and incorporate them as part of each team's appointment.

DIGITAL CONSTRUCTION LEAD

To ensure that all necessary information requirements are fulfilled, the lead appointed party should create a comprehensive set of exchange information requirements. These should not only consider what the appointing party needs but also include any extra requirements set by the lead appointed party, which the task team must follow as part of the appointment.

This step is important because the appointing party's needs might not encompass everything required for the long-term management of information and data analytics by the lead appointed party. Therefore, the lead appointed party should carefully evaluate all requirements and check that their needs are addressed within all the different appointments with each task team.

Establish the task information delivery plan (Clause 5.4.4)

The task information delivery plan (TIDP) needs to be produced and kept updated throughout the duration of the appointment.

TASK TEAM

Task teams are each responsible for developing and maintaining a task information delivery plan (TIDP) that aligns with their scope of work and the detailed responsibility matrix. Teams should use the project template rather than creating their own format for the TIDP. In crafting the TIDP, consider such factors as information delivery milestones, requirements from the lead appointed party, shared resources and the time needed to produce, coordinate, review and approve information.

The TIDP should list such details as the name and title of each information container, any dependencies, the level of information need, the estimated production duration, the author responsible and delivery milestones. All documents in the TIDP must follow the project naming protocol and their delivery dates must correspond to the project programme.

DESIGN MANAGER AND TECHNICAL SERVICES MANAGER

As part of their roles in overseeing the project's design and technical services, the design manager and technical services manager are responsible for reviewing the TIDP submitted by each task team. This review aims to ensure that the TIDP's content meets the project's needs and expectations.

During the review, particular attention must be paid to the number of suggested drawings for each package. The managers must verify that the information accurately reflects the project's requirements and that the drawings are formatted correctly. Such attention to detail is crucial for ensuring the TIDP's accuracy, thereby preventing its rejection when uploaded to the common data environment and avoiding delays in the overall approval process.

In addition to the responsibilities already outlined, the design manager and technical services manager have another essential task: to make sure that each task team understands the level of information expected for delivery at each milestone.

It's important to distinguish between the 'first issue date' and the 'delivery date' of a document. The 'first issue date' is the initial upload date of the document to the common data environment, while the 'delivery date' is when the document is issued for review and authorisation by the lead appointed party. The terminology used to describe these concepts may vary within your business, so it's essential to clarify these terms to ensure smooth communication.

While this distinction may seem minor, it is relevant for ensuring that each task team submits the required information with the appropriate quality content by the delivery date. A lack of clarity could lead to submissions of incomplete or insufficiently detailed information, merely to meet the deadlines, causing project delays.

By clarifying these dates and the information expected at each milestone, the design manager and technical services manager can make sure that the rest of the delivery team receives the necessary information on time and to the expected standard. This focus on detail and effective communication between the task teams will help to avert any unnecessary delays and contribute to the project's success.

COMMERCIAL

The TIDP should be completed and issued during the tender stage by each task team, and it should be included in their appointments.

This method has proven effective in preventing problems, such as late delivery of the TIDP and unwillingness to complete it punctually. On some occasions, the TIDP has been misused as a drawing register rather than serving its intended purpose. By formally incorporating the TIDP in the appointments, a clear expectation is set for its timely completion and delivery. This also acts as a valuable commercial tool for addressing any issues of delays or poor performance by any party involved. Including the TIDP in appointments further clarifies roles and responsibilities, helping to avoid misunderstandings or disputes over required information and deadlines, ultimately contributing to the project's success.

Establish the master information delivery plan (Clause 5.4.5)

Each task information delivery plan (TIDP) produced by each task team will be consolidated by the lead appointed party to form the master information delivery plan (MIDP).

141

DESIGN MANAGER

The design manager has the responsibility of managing and maintaining the master information delivery plan (MIDP) by integrating individual task information delivery plans (TIDPs), according to the programme's requirements. To manage this information efficiently and evaluate the impact of any changes on individual TIDPs within the overall programme, it's advisable to use digital tools. These tools can help maintain the accuracy and consistency of the MIDP while also offering real-time updates on changes to TIDPs. This approach enables effective communication between project stakeholders and helps ensure that the project stays on course.

Complete the lead appointed party's appointment documents (Clause 5.4.6)

While this is the responsibility of the appointing party, the lead appointed party must be aware of this clause and should ensure that the appropriate documents have been reviewed and included as part of their appointment with the appointing party.

PRECONSTRUCTION MANAGER

To ensure the success of the project, the preconstruction manager must carefully review and manage the key documents that need to be included in the appointment. These documents should also be managed through change control for the duration of the appointment.

By meticulously reviewing and managing these relevant documents, the preconstruction manager can make sure that the information in the appointments is consistent with the agreed approach. Additionally, any changes that are made will be properly documented and controlled.

Complete the appointed party's appointment documents (Clause 5.4.7)

The lead appointed party is responsible for incorporating the appropriate documents in each task team's appointment. They must also manage these documents through change control for the duration of the appointment.

COMMERCIAL

It's fundamental for the commercial team to fully grasp the extent of compliance and participation needed from each task team during the BIM implementation of the project. Different task teams may require different documentation for their appointments. A thorough understanding of this process allows the commercial team to provide relevant information during the tender stage and to make appropriate appointments with the various parties involved. This is crucial for ensuring that BIM requirements are met throughout the project's lifecycle.

Mobilisation (Clause 5.5)

A summary checklist of the mobilisation team actions is given in Table 6.3.

Mobilise resources (Clause 5.5.1)

The lead appointed party is responsible for deploying the resources in accordance with the mobilisation plan submitted at tender stage.

INFORMATION MANAGER

The information manager has a key role in collaborating with the appointing party to establish a naming convention and develop a naming protocol. This protocol sets a standard naming system for all project information. All team members will use this system to name any information included in their TIDPs. This ensures consistent collaboration and simplifies the search for information within the common data environment (CDE).

Table 6.3 Checklist: mobilisation team actions

Mobilise resources (Clause 5.5.1)

Information manager
- ☑ Establish a naming convention and protocol.
- ☑ Ensure all project information adheres to naming convention.
- ☑ Define the folder structure as needed, and specify the required metadata.
- ☑ Implement appropriate security measures for the CDE.

Design manager and operations lead
- ☑ Initiate first design team meeting.
- ☑ Assemble task teams.
- ☑ Emphasise collaboration among team members.
- ☑ Encourage attendance at BIM and CDE training sessions.

Digital construction lead
- ☑ Equip team members with necessary skills and knowledge.
- ☑ Provide project-specific training sessions.
- ☑ Make training sessions mandatory.

Mobilise information technology (Clause 5.5.2)

Information manager
- ☑ Establish a common data environment (CDE).
- ☑ Configure CDE workflows and access permissions.
- ☑ Provide CDE training for team members.
- ☑ Act as the first point of contact for CDE-related issues.
- ☑ Define workflows and requests for information (RFIs).

Design manager and technical services manager
- ☑ Collaborate with information manager to establish CDE workflows.
- ☑ Ensure effective communication about workflows.

Digital construction lead
- ☑ Set up and provide access to digital tools.
- ☑ Organise training sessions for digital tools.
- ☑ Monitor effectiveness of digital tools.
- ☑ Solicit feedback on digital tools.

Test the project's information production methods and procedures (Clause 5.5.3)

Information manager
- ☑ Test information exchanges.
- ☑ Balance between security and efficiency.
- ☑ Provide support and training to teams.

Design manager and technical services manager
- ☑ Ensure clear understanding of information production methods.
- ☑ Ensure effective team communication.
- ☑ Verify accuracy of survey information.

Digital construction lead
- ☑ Ensure consistency in file formats.
- ☑ Provide necessary resources for surveys.
- ☑ Identify potential issues early in the project.

After agreeing on the naming convention and protocol, the information manager ensures that all project information meet the agreed standards. This minimises confusion and errors stemming from inconsistent or conflicting naming systems.

Another vital task for the information manager is to define a well-structured folder system, if required, and the metadata to be used. This facilitates data search and ensures that essential project information is readily available to team members. Additionally, robust security measures will be implemented to restrict access to authorised team members only.

DESIGN MANAGER AND OPERATIONS LEAD
The design manager and operations lead kick off the project by initiating the first design team meeting. During this meeting, task teams are assembled and the importance of collaboration is stressed. These leaders create a collaborative environment to ensure compliance with BIM project requirements throughout the project's lifecycle.

They should also encourage team members to attend relevant training sessions offered by the digital construction lead or information manager. These sessions aim to clarify BIM expectations, the correct use of the CDE and information approval workflows. Failure to attend these training sessions can lead to project inefficiencies and errors. Therefore, it's crucial that all team members understand the importance of this training and the need to comply with BIM project requirements.

DIGITAL CONSTRUCTION LEAD
The digital construction lead will ensure that all task team members are adequately skilled and knowledgeable to meet the BIM expectations and project requirements. Although these expectations are outlined during the tender stage, team composition may have changed, requiring updated, tailored training.

Training sessions must cover the full scope of the project, from specific requirements to roles and responsibilities throughout the project. These sessions should also include lessons learned from previous projects, serving as a valuable refresher course on the BIM process.

Attendance at these training sessions is mandatory for all internal team members, regardless of their prior experience. This ensures that the team is cohesive and collaboratively focused on achieving project goals. By making these training sessions obligatory and underscoring their importance, the digital construction lead fosters a collaborative and efficient work environment, contributing to the project's successful execution.

Mobilise information technology (Clause 5.5.2)
The lead appointed party shall mobilise the information technology in accordance with the mobilisation plan submitted at tender stage.

INFORMATION MANAGER
The information manager is tasked with setting up and managing the project's common data environment (CDE), a crucial component of a successful project. The CDE serves as the authoritative source of all project-related information, offering a central repository for data storage, sharing and collaboration among team members.

To ensure effective use of the CDE throughout the project, the information manager will collaborate with the design manager, technical services manager and appointing party. Together, they will configure workflows and set access permissions for all project team members. The information manager will also provide relevant training and access to the CDE for both internal and external team members, including training refreshers and onboarding for new users.

The information management team acts as the first point of contact for any queries or guidance needed regarding the CDE. They will facilitate rapid access to the CDE for project team members and revoke access when a member leaves the project. The team will also define workflows, establish requests for information (RFIs) and design development processes, and make shared resources, such as point cloud surveys, available to team members. Additionally, they are responsible for testing information exchanges and ensuring that the data delivered meet project requirements

DIGITAL CONSTRUCTION LEAD

The digital construction lead has a primary responsibility for setting up and granting access to various digital tools essential for design coordination throughout the project. This includes, but is not limited to, model viewers, 360° site captures point cloud data and the application of AI and data analytics.

To equip project team members with the skills needed to use these digital tools, the digital construction lead will organise training sessions or workshops tailored to the specific needs of each task team member.

The digital construction lead is also responsible for managing the deployment of these digital tools. This entails overseeing tool usage, monitoring effectiveness and making any necessary adjustments. Feedback from project team members should be actively sought to identify areas for improvement.

DESIGN MANAGER AND TECHNICAL SERVICES MANAGER

The design manager and technical services manager must closely collaborate with the information manager to create project-specific workflows within the CDE. These workflows should align with contractual requirements and receive agreement from all project stakeholders.

Regular dialogue between the design manager, technical services manager and information manager is necessary for ensuring that the workflows function as intended. This collaborative approach helps to promptly address any issues and minimises risks such as delays or errors, which could adversely affect the project's progress.

Test the project's information production methods and procedures (Clause 5.5.3)

The lead appointed party shall test the project's information production methods and procedures and ensure that they are understood by all members of the delivery team, in accordance with the mobilisation plan submitted at tender stage.

INFORMATION MANAGER

The information manager is in charge of testing the system to ensure that all involved parties can upload and download information while adhering to the agreed workflow. The role also includes offering clarification and support to the teams as needed.

Testing the exchange of information is crucial for confirming that the system is both secure and efficient. While security is paramount, an excessively complex process that hampers quick access to information is undesirable. Hence, the information manager must find a balance between security and efficiency with the appointing party to ensure that the system operates seamlessly.

The information manager is also tasked with supporting the whole project team. This support may come in the form of training sessions or tutorials aimed at helping team members grasp the workflow and use the system effectively. Availability to address questions or solve issues that may arise is also essential.

DESIGN MANAGER AND TECHNICAL SERVICES MANAGER

The design manager and technical services manager are tasked with ensuring that the delivery team clearly understand the methods and procedures for producing information. Effective communication with the team will ensure that everyone is working together and that there is no scope for confusion or misinterpretation.

A critical aspect of the design manager's and technical services manager's roles is communication to ensure the project runs smoothly. They must inform all relevant parties about the processes involved in producing the project's information.

Accurate survey information is crucial for developing a design that fulfils the project's needs. Thus, it falls under the design manager's and technical services manager's purview to make sure that survey information is complete, accurate and includes all necessary details. This supports the design team in proceeding with the work and meeting the deadlines outlined in the TIDP.

DIGITAL CONSTRUCTION LEAD

The digital construction lead is responsible for ensuring that the project's different file formats are consistent, share the same coordinates and satisfy the project requirements.

In addition, the digital construction lead should provide the necessary resources to the delivery team, including tools and a survey strategy for validating and verifying design information or assessing the site's existing condition. Conducting these surveys allows the task teams to identify any potential challenges early in the project, thereby helping to avert delays and cost overruns.

Collaborative production of information (Clause 5.6)

A summary checklist of the production of information team actions is given in Table 6.4.

Check availability of reference information and shared resources (Clause 5.6.1)

TASK TEAMS

Task teams must each ensure that they have access to relevant information and shared resources in the common data environment before generating any information. If access is not available, the team should inform the lead appointed party and evaluate the potential delay this could cause to their TIDP.

INFORMATION MANAGER

The responsibility of the information manager is to guarantee that the task team members listed in the project directory can access the common data environment and that the information on shared resources is available to the teams.

DESIGN MANAGER, TECHNICAL SERVICES MANAGER AND OPERATIONS LEAD

The design manager and technical services manager will confirm with each task team that they comprehend the information available and handle any enquiries. The MIDP must be updated regularly to reflect the latest TIDP from each task team and any project changes. Additionally, the design manager is accountable for maintaining the project directory to enable the information manager to provide access to the appropriate personnel.

Table 6.4 Checklist: production of information team actions

Check availability of reference information and shared resources (Clause 5.6.1)
Task teams
☑ Check for access to relevant information and shared resources.
☑ Inform the lead appointed party if access to resources is unavailable.
☑ Assess potential impact on TIDP.

Information manager
☑ Ensure that task team members have access to the common data environment.
☑ Make sure information on shared resources is visible to teams.

Design manager, technical services manager and operations lead
☑ Confirm that the design team comprehends the information.
☑ Handle any enquiries.
☑ Regularly update MIDP.
☑ Maintain the project directory.

Generate information (Clause 5.6.2)
Task teams
☑ Generate project information according to TIDP and project information standards.
☑ Discuss any deviations with the design manager.
☑ Coordinate with other task teams for information and spatial coordination.
☑ Collaborate to resolve coordination issues or inform the lead appointed party.

Undertake quality assurance check (Clause 5.6.3)
Task teams
☑ Conduct a quality assurance check on the information container.
☑ Follow the project's information standard during the check.
☑ Mark the information container as checked if successful.
☑ Reject the container and inform the author if unsuccessful.

Review information and approve for sharing (Clause 5.6.4)
Task teams
☑ Review the information within the container according to project's methods.
☑ Consider lead appointed party's requirements and necessary coordination.
☑ Assign a suitability code and approve for sharing if checks are successful.
☑ Record reason and reject the container if unsuccessful.

Information model review (Clause 5.6.5)
Task teams
☑ Review the information model for consistency.
☑ Adhere to the project's methods and consider information requirements.
☑ Regularly review the design model.
☑ Participate in coordination meetings.
☑ Update the information model until ready for lead review.

Design manager and technical services manager
☑ Ensure that task teams work collaboratively.
☑ Ensure timely resolution of design errors.
☑ Supervise the design development process.
☑ Take ownership of accurate and complete drawings.

Generate information (Clause 5.6.2)

TASK TEAMS

The various task teams generate project information in accordance with the agreed TIDP and project information standards. Any deviation from this will be discussed with the design manager. The design manager addresses design issues and coordinates with task teams to deliver the required level of information. Furthermore, the project necessitates information coordination and spatial coordination of geometrical models among task teams using the common data environment. Task teams must collaborate to resolve coordination issues or inform the lead appointed party if no solution is found.

Undertake quality assurance check (Clause 5.6.3)

TASK TEAMS

Each task team will conduct a quality assurance check of the information they have produced, in accordance with the project's information production methods and procedures. This initial check is to ensure that the name and metadata of the information are correct, before a thorough review is conducted.

The task teams will review the information container to ensure that it meets the project's standards. After the review, the task team will do one of two things: either approve the container, mark it as checked and record the outcome or reject it and notify the author about the corrections needed.

Review information and approve for sharing (Clause 5.6.4)

TASK TEAMS

Each task team will conduct a quality assurance check of the information produced, following the project's established methods and procedures. This check will be completed before sharing the information within the project's common data environment. This second check is to ensure that everything aligns with the exchange information requirements and the task information delivery plan.

Task teams need to consider the lead appointed party's information requirements, the level of information need and information necessary for coordination by other teams. After the review, if the internal checks are successful, the teams will assign a suitability code and approve the container for sharing. However, if the review is unsuccessful, they will record the reason, specify any necessary amendments and reject the container.

Information model review (Clause 5.6.5)

The delivery team will assess the information model, adhering to the project's information production methods and procedures. This clause will be repeated as necessary until the information model is ready to be authorised by the lead appointed party.

TASK TEAMS

The task teams must adhere to the project's information production methods and procedures. It is also crucial to take into account the information needs of the appointing party, as well as the requirements outlined in the master information delivery plan.

Regular reviews of the design model are essential. Coordination meetings should be held throughout the design development phase with the aim of identifying and resolving design issues in a prioritised manner. The frequency of these model reviews will vary, depending on the project's current status and overall timeline.

Thanks to new solutions in the market, these meetings no longer need to be as frequent as in the past. The design team meetings can focus on key coordination issues. The information model will be updated continuously until it is ready for review and authorisation by the lead appointed party.

DESIGN MANAGER AND TECHNICAL SERVICES MANAGER

At this stage, the design manager and technical services manager ensure that all task teams collaborate effectively and share essential information by the agreed deadlines. The project's success hinges on the teams' abilities to spot and promptly rectify design errors, contributing to significant progress in the design process. It's not merely about identifying problems; rapid resolution is equally important. This is crucial because design errors are often where task teams fall behind schedule. Therefore, ongoing oversight by the design manager and technical services manager is essential.

When task teams collaborate effectively to produce a well-coordinated, high-quality design model, this leads to the delivery of excellent design information. This is key for the project's success. Both the design manager and the technical services manager are responsible for ensuring that all parties identify and correct design coordination errors in a timely fashion, to prevent any negative impacts on the programme.

A critical aspect of this phase is making sure that any agreed design changes are incorporated into the model and reflected in the construction drawings. Drawings with unresolved issues should not be authorised for construction, as these errors will likely carry over to the building site. Ensuring that drawings are both accurate and complete is a responsibility that falls squarely on the shoulders of the design manager and the technical services manager.

Information model delivery (Clause 5.7)

A summary checklist of the information model delivery team actions is given in Table 6.5.

Submit the information model for lead appointed party authorisation (Clause 5.7.1)

TASK TEAM

The task team must submit project information on time to the common data environment for review and authorisation by the lead appointed party.

Review and authorise the information model (Clause 5.7.2)

The lead appointed party will review the information provided by each task team, following the project's methods and procedures for information production. If the review is successful, the lead appointed party will authorise the information for acceptance by the appointing party, using the common data environment.

INFORMATION MANAGER

The information manager plays a key role in upholding the quality of the project's common data environment (CDE) and has the main task of ensuring that all information uploaded to the CDE complies with the project's quality guidelines. This involves an initial quality check to confirm that the information aligns with the project's naming protocol. The information manager also ensures that correct metadata, such as status, revision, classification and a unique name, are applied to prevent duplication.

Table 6.5 Checklist: information model delivery team actions

Submit information model for lead appointed party authorisation (Clause 5.7.1)
Task team
☑ Submit project information to the common data environment (CDE) on time.

Review and authorise the information model (Clause 5.7.2)
Information manager
☑ Conduct an initial quality check for compliance with the project's naming protocol and metadata.
☑ Ensure deadlines for uploading information to the CDE are met, following the TIDP (task information delivery plan).
☑ Produce periodic reports detailing outstanding CDE actions for internal and external teams.

Design manager and technical services manager
☑ Review and validate the quality of the data provided by task teams.
☑ Ensure that the information complies with the master information delivery plan (MIDP) and the level of information need.
☑ Oversee the distribution of relevant information to appropriate task teams for review and comments.

Digital construction lead
☑ Keep track of the project's progress and address outstanding actions.
☑ Conduct appropriate reports at each stage of the project.
☑ Verify that the data provided are accurate and have been validated by the appropriate team members.

Submit information model for appointing party acceptance (Clause 5.7.3)
Task team
☑ Submit information with metadata for appointing party acceptance in the common data environment (CDE).

Review and accept the information model (Clause 5.7.4)
Information manager
☑ Record appointing party's outcome in the report.

Design manager and technical services manager
☑ Review appointing party's outcome.
☑ Identify required actions from task teams.
☑ Track and monitor updates from task teams.
☑ Ensure that new information satisfies appointing party's requirements

---

For the seamless flow of information to other teams, it's necessary for the information manager to conduct the initial quality assurance swiftly. This expedites the availability of information for other teams. A well-structured task information delivery plan (TIDP) minimises the risk of failing the initial quality check, thus cutting down time spent on resubmissions and approvals.

The TIDP outlines deadlines for uploading information to the CDE, and the information manager must make sure these deadlines are met. Any delays can disrupt the work of other teams, making it essential to stick to the schedule in the TIDP.

To monitor project progress, the information manager should generate regular reports, detailing any pending CDE actions required from both internal and external teams. These reports are key for keeping everyone informed about the project's status and for addressing any delays or issues promptly.

DESIGN MANAGER AND TECHNICAL SERVICES MANAGER

The design manager and technical services manager have to ensure that the information delivered by each task team meets the project quality standards. They review this information for alignment with the master information delivery plan (MIDP) and the level of information need (LOIN). They also verify the data quality as specified in the BIM Execution Plan, in collaboration with the digital construction lead.

To streamline the approval workflow, the design manager and technical services manager need to be actively involved in the information management process. After the information management team completes the initial quality check, the relevant data should be distributed to the respective task teams for review and comments. This avoids information overload and ensures that the approval process progresses in a timely manner. Both managers are responsible for overseeing this procedure and for ensuring that all required approvals are secured before progressing. Partial acceptance of the information container can lead to coordination issues, so it's advisable to either approve or reject the information in its entirety.

DIGITAL CONSTRUCTION LEAD

The digital construction lead is responsible for monitoring the project's progress and making sure that any outstanding tasks are promptly addressed. They must prepare reports at different stages of the project and, when necessary, highlight any pending actions and to ensure that the information meets the client's BIM expectations. A crucial part of this role is to validate the accuracy of the data, making sure it is reliable for use by the delivery team.

The digital construction lead identifies any errors or omissions in the information, allowing for quick corrective action. Regular reporting and verification of the project progress enables the delivery team to identify and address any issues promptly, ensuring the project is completed on time and within budget, and meets client expectations.

Submit the information model for appointing party acceptance (Clause 5.7.3)

TASK TEAM

Task teams are each required to submit their information, with the correct metadata, in the common data environment for the review and acceptance of the appointing party.

Review and accept the information model (Clause 5.7.4)

While the appointing party is responsible for this review, the lead appointed party must closely monitor the results of the appointing party's review of the information. The lead appointed party needs to know whether the information has been accepted or rejected in order to direct the task teams to make any required amendments.

INFORMATION MANAGER

The information manager is responsible for recording the outcome of the appointing party's review in the report.

151

DESIGN MANAGER AND TECHNICAL SERVICES MANAGER

The design manager and technical services manager will monitor the outcome of the review. If any actions are required from the task teams, they will track and monitor updates to ensure that the new information meets the appointing party's requirements.

Execution of construction works

A summary checklist of the execution of the construction works is given in Table 6.6.

While there are no specific clauses from the ISO 19650-2 related to the execution of construction works, it is important to discuss some key considerations. The focus often falls on information management during the design phase, but it's equally crucial for on-site teams to use the provided information accurately and maintain its integrity throughout the construction phase. This aspect can sometimes be overlooked.

Table 6.6 Checklist: execution of the construction works

Execution of the construction works

All teams
- ☑ Attend training sessions and workshops to understand the BIM process.
- ☑ Follow procedures and guidelines in the project's BIM Execution Plan.
- ☑ Regularly update project status and attend status meetings.

Operations team
- ☑ Use up-to-date digital construction information.
- ☑ Avoid printing drawings; use digital tools.
- ☑ Regularly review the model on site with stakeholders.
- ☑ Provide access to the model for all contractors.
- ☑ Ensure execution of work corresponds to latest accepted construction information.
- ☑ Verify data provided by contractors with design manager and technical services manager.
- ☑ Ensure O&M information complies with customer requirements.

Information manager
- ☑ Gather O&M data with the site team.
- ☑ Cross-check O&M information for consistency with project requirements.
- ☑ Liaise with the appointing party for information transfer to common data environment.

Commercial team
- ☑ Verify that subcontractors have fulfilled their contractual BIM requirements.
- ☑ Implement a verification process for submitted information.

Task team
- ☑ Revise design information based on discoveries made during construction.
- ☑ Update design information to reflect changes.
- ☑ Ensure that design changes follow the appropriate change control process.

Digital construction lead
- ☑ Support project teams in accessing the information model.
- ☑ Encourage use of the model viewer and the use of technology to support on-site verification.
- ☑ Maintain a visible and accessible presence on site.

ALL TEAMS

It's essential to stress the importance of adopting the correct approach to information management during the construction stage. During this phase, pressures to resolve conflicts on site can lead to shortcuts that deviate from best practices, resulting in issues at a later stage. One way to mitigate this is by conducting training sessions and workshops. These sessions can educate all team members about proper procedures and the importance of following the appropriate process, as well as embracing the digital tools available for the project.

The project's success is closely tied to how well the delivery teams use the common data environment (CDE) and information models during construction. Therefore, it's fundamental to monitor their progress and provide the necessary support. Regular status reports and meetings can serve as platforms to discuss any outstanding actions, resolve issues and offer additional training where needed.

OPERATIONS TEAM

The successful delivery of a construction project depends heavily on effective collaboration and communication between the operations team and other stakeholders. Accurate information management during the construction stage is also vital. The operations team should use the latest accepted construction information to maintain high-quality work and avoid errors on site. One way to do this is to use digital tools instead of printed drawings. This approach minimises the risk of using outdated information.

Additionally. the operations team should regularly review the model on site with different stakeholders encouraging the model's use by all parties. This approach can help ensure that subcontractors have a correct understanding of the works to be conducted on the project.

The operations team should confirm that the execution of the works aligns with the latest accepted construction information, taking accountability for ensuring that the asset data required as part of the client requirements has been provided on time by the different task teams and validated by the digital construction lead. To achieve this, the operations team should work collaboratively with the design manager and technical services manager to verify the asset data provided by subcontractors.

Furthermore, the operations team is tasked with ensuring that any design alterations agreed on during the construction stage are processed through the agreed change control. It is crucial that information updates reflecting these changes are made before conducting the works on site. This approach will support the accurate delivery of the works and ensure that the information provided to the appointing party accurately reflects the completed works.

Lastly, the operations team should also be accountable for verifying that the operations and maintenance (O&M) documentation complies with the client requirements. This involves collaborating with the information manager and the task teams to collect all the requisite information, ensuring it meets the project requirements and contains the appropriate metadata. By adhering to these guidelines, the operations team can ensure the successful delivery of a construction project that meets the client's requirements and expectations.

INFORMATION MANAGER

The information manager works together with the site team to gather all the relevant operations and maintenance documentation and cross-checks the information to ensure that it is consistent with the project requirements and contains accurate metadata. Additionally, the information manager liaises with the appointing party to ensure that the necessary information is transferred to their common data environment (CDE) or other relevant platform to be utilised for managing the information throughout the operational phase of the asset.

Effective collaboration between the lead appointed party and the appointing party is essential for ensuring that all O&M documentation is managed and transferred in a timely manner and at the appropriate level of quality. This facilitates the accurate handover of information, allowing the appointing party to access the relevant data as quickly as possible.

COMMERCIAL

It is important to verify that all task teams have fulfilled their contractual BIM project requirements before their accounts are closed. This step is necessary for ensuring that all relevant information is provided and that the project is delivered successfully. To achieve this, a thorough verification process must be implemented to confirm that all information has been submitted and with the quality required.

Select subcontractors known for quality and timeliness. If a subcontractor misses a deadline for providing essential information, instead of using retention, consider employing diligent contract management. The Construction Leadership Council (CLC) and NEC (2022) advise transitioning away from retentions by 2025, recommending alternative strategies to foster high-quality work and ensure that project deadlines are met. Maintain open communication to address and resolve any issues promptly.

TASK TEAM

It is crucial to keep design information current and to revise it based on any changes during construction. These adjustments must follow a stringent change control process to ensure proper documentation review and must be authorised before execution of the works on site.

DIGITAL CONSTRUCTION LEAD

The digital construction lead will support the delivery team to access and use the information model effectively on site. This should involve promoting the use of mobile devices to easily check completed works while on location.

To make this a success, the digital construction team needs to be visible and easily approachable on site. This will help remove any obstacles and forge a connection between the digital construction procedures and the delivery team, whether they are internal or external. A visible presence and the right support from the digital construction lead will help to ensure the successful adoption and use of BIM throughout the construction phase.

Project close-out (Clause 5.8)

A summary checklist of the project close-out team actions is given in Table 6.7.

Archive the project information model (Clause 5.8.1)

Like the appointing party, the lead appointed party is responsible for storing the information containers after handover in their common data environment and creating a backup copy to ensure that the information is securely stored and readily accessible for future reference if required.

OPERATIONS LEAD

The operations lead must collaborate closely with the appointing party to guarantee that all project information satisfies the requirements. It's essential to gather feedback on the client's experience with the digital construction process as part of the client satisfaction questionnaire. This feedback will be shared with the digital construction lead to identify successes and areas for improvement. In this way, the team can continually improve the digital construction process and better meet the

Table 6.7 Checklist: project close-out team actions

Archive the project information model (Clause 5.8.1)

Operations lead

☑ Collaborate with the appointing party to ensure that project information meets requirements.

☑ Gather feedback on the customer's experience of the digital construction process.

☑ Share this feedback with the digital construction lead for analysis.

☑ Continually improve the digital construction process based on feedback to better meet customer needs.

☑ Maintain open communication with the customer to ensure a successful project outcome.

Information manager

☑ Archive all project information for future access.

Capture lessons learned for future projects (Clause 5.8.2)

Digital construction lead

☑ Capture feedback and experiences from stakeholders involved in the project.

☑ Use the feedback to improve project delivery processes and documentation.

☑ Identify updates required to the process and documentation for efficient use of digital tools.

☑ Prepare a case study outlining the process and experiences, both positive and negative.

☑ Prepare any marketing content necessary to communicate the project's success to internal and external teams.

☑ Capture the performance data of the different task teams.

☑ Incorporate performance data into a database for data analytics to support business decisions.

---

client's needs. Open communication with the client and the collection of feedback are critical to ensuring a successful project outcome and fostering strong relationships with clients.

INFORMATION MANAGER

The information manager is responsible for archiving project information with the relevant metadata to ensure that it is accessible for future requirements.

Capture lessons learned for future projects (Clause 5.8.2)

Just like the appointing party, the lead appointed party should collaborate with all project team members to document lessons learned at each stage of the project. These should be recorded for reference in future projects.

DIGITAL CONSTRUCTION LEAD

The digital construction lead holds several responsibilities, aimed at ensuring the project's success and capturing insights for future improvements. One key role is to gather feedback and experiences from various stakeholders involved in the project. This information is used to refine the project delivery process and update documentation for future projects, ensuring efficient use of digital tools.

Another responsibility involves preparing a case study that outlines the project's processes, highlighting both positive and negative experiences. This serves as a valuable resource for others looking to understand and learn from the project. Additionally, the digital construction lead is tasked with creating any marketing content needed to share the project's success with both internal and external teams.

It's also crucial to monitor the performance of different task teams, to inform broader business decisions and for consideration when selecting teams for future projects. The digital construction lead compiles a database of this information, providing a solid foundation for data analytics to support business decisions.

6.3. Appointing party strategy

The appointing party is the client or the party responsible for managing information on behalf of the client. In the past, I have encountered situations where the tender information provided by the appointing party includes a request for implementing BIM Level 2 without providing further details on the appointing party's expectations or the data required to support the operations and maintenance stage. Also, sometimes the appointing party's BIM documentation is based on templates with a great many data requirements and other content that might not be relevant to the project or business needs. This could mean that the appointing party is simply ticking a box without a real interest in BIM or might not fully understand the process and what needs to be considered to make the BIM process valuable in the long run.

To ensure successful implementation of the BIM process, it is crucial to discuss this with the appointing party at the earliest stage possible and effectively clarify the requirements for the project. This is also an opportunity to provide guidance and support, if the appointing party is not very experienced with the BIM process. This support can include workshops, examples from previous projects and advice on best practice. Providing this kind of support can help the appointing party feel more confident about using BIM and maximise its benefits for the project. These discussions not only dispel common misconceptions about additional time and costs but also provide an opportunity to address information and data management requirements, information flow and responsibilities, and collaborative working methods, resulting in smoother project execution and greater overall success.

Therefore, to ensure the successful implementation of BIM, the appointing party must follow the proper process and provide all required documentation at the tender stage. It is also necessary to adopt a business strategy approach to BIM, which involves having the necessary resources and expertise, as well as a plan for integrating it in the company's overall operations.

A key benefit is the ability to make informed design decisions and have a better understanding of the proposed design, by having all project information integrated in a comprehensive digital model. This improved visual representation of the design and data can help the client make informed decisions about the design, materials and use of spaces.

It is important for the appointing party to understand that the benefits of BIM extend beyond the design and construction phase. The BIM process can enhance the operations and maintenance of the completed project through accurate information management and comprehensive asset data collection. This accurate and easily accessible information will prove invaluable for the facility management team.

Furthermore, by utilising a comprehensive and precise digital representation of the asset, facility managers can easily pinpoint the location of assets and promptly resolve any potential problems, based on the correct information. This enables them to make well-informed decisions concerning asset maintenance and improvements, leading to reduced expenses and improved overall performance in the long run.

Appointing party information management function

The information management function is an indispensable component of any project and plays a vital role in ensuring the effective management of information throughout its lifecycle. Hence, when additional support is required to conduct the information management function, it is critical that the appointing party nominates an individual within the organisation or an external consultant to conduct this function. I will refer to the person who provides this support as the appointing party's information manager, but it should be noted that the functions and responsibilities are not unique to specific job titles or professionals. It is important to involve the information manager early in the strategic definition stage, to define goals and develop a comprehensive strategy for managing the portfolio information.

The appointing party's information manager can provide valuable insight into the project and help the appointing party identify long-term objectives, determine the data that will be required throughout the project and develop a robust plan for managing the information effectively.

It is also essential for the appointing party's information manager to remain actively engaged in the project, to ensure that the delivery team is producing the necessary information according to the client's expectations and to be available to answer any questions or provide clarification as needed.

The appointing party's information manager should review the tender response of the prospective lead appointed party, including the pre-appointment BIM Execution Plan (BEP), capability and capacity assessment summary, mobilisation plan and information delivery risk assessment. Additionally, the information manager should assess and review the compliance of the appointment documents for the lead appointed party, including the appointing party's exchange information requirements, the project's information standard and the project's information protocol, including any agreed additions or amendments, the delivery team's BEP and the delivery team's MIDP. This review process is essential to ensure that all parties involved in the project are working towards the same objectives and that information and data are being appropriately managed and delivered.

Collaboration between the appointing party's information manager and the lead appointed party is essential to ensure that the project is proceeding as planned and that the appointing party's needs are being met. The appointing party's information manager should provide feedback to the lead appointed party during the project, to enable any necessary adjustments to meet the client's needs. In this way, the project can progress smoothly and meet the client's expectations, leading to a successful outcome.

The responsibilities held by the appointing party's information manager cannot be underestimated when it comes to the proficient handling of information and data throughout a project's lifecycle.

The specific duties assigned to this role might undergo modifications, based on the varying needs of the appointing party and the unique demands of the project itself. However, the following aspects remain paramount and should be given considerable thought.

- *Appreciation of the appointing party's needs*. Profoundly understanding the appointing party's objectives and requirements is indispensable. This ensures that the overall information management strategy meets the appointing party's expectations and is bespoke to the project goals.

- *Understanding of design, construction and operations*. A robust understanding of the processes involved in design, construction and operations is necessary. This involves supporting the validation of the quality and accuracy of information produced to serve various project stages effectively.

157

- *Knowledge of BIM standards and protocols.* Proficiency in standards and protocols is crucial for maintaining compliance consistently throughout the project, ensuring alignment with industry best practices and setting quality benchmarks.
- *Assisting with documentation.* Active involvement in the production of necessary documentation is essential. This documentation should be produced in accordance with the appointing party's business strategy and define project-level requirements clearly, promoting transparency and understanding among stakeholders.
- *Guidance on tool selection.* Advising on the choice of suitable tools that offer robust support to the appointing party's needs throughout the project lifecycle is key to improving efficiency and fostering better project outcomes.
- *Integrating workflow.* Facilitating the incorporation of the BIM process in the organisation's workflow and ensuring its effective utilisation throughout the project lifecycle contributes significantly to the project's success.
- *Training and collaboration.* Providing comprehensive internal training and ongoing collaboration to external parties enhances the team's understanding and clarifies requirements during project delivery, ensuring a unified approach towards shared goals.
- *Advising on asset management.* Offering advice on using the produced information for effective asset management is crucial. This includes the creation of digital twins for monitoring and maintaining buildings and infrastructure, enhancing long-term operational efficiency and sustainability.
- *Ensuring data accuracy.* Rigorously validating the accuracy and completeness of information and data provided by the delivery team is necessary to ensure decisions are based on reliable and accurate data, preventing potential inaccuracies that can significantly impact the project's outcome.
- *Capturing lessons learned.* Meticulously documenting lessons learned from the project and applying this invaluable knowledge to future projects promotes continuous improvement and growth.
- *Security and confidentiality.* A thorough understanding of the project's security requirements and the importance of maintaining confidentiality is crucial for protecting and appropriately managing sensitive information.

Hierarchy of information requirements

Effective information management is not just about regulatory compliance; it also contributes to the efficient execution and maintenance of projects. For the appointing party, incorporating information management in both project-specific considerations and overarching business strategies is vital.

This requires appointing parties to possess a thorough understanding of the information needed to accomplish their objectives across the asset's lifecycle and to clearly communicate these requirements to all project stakeholders. Such information could range from maintaining an asset register, satisfying compliance and regulatory obligations and managing risks to aiding in business decision making.

For informed decision making, the appointing party must have access to a variety of information. This includes data on the asset's intended capacity and usage, security and surveillance needs, renovation needs, forecast and actual impacts, operational needs, maintenance and repair requirements and costs related to replacement, decommissioning and disposal. This pool of information enables better anticipation and planning of costs, effective asset management and adherence to regulatory and compliance standards.

To ensure that the information requested for a project aligns with the broader business strategy, it should remain consistent and in line with the strategy. The appointing party can refer to the 'information requirements hierarchy' included in ISO 19650-1 (Figure 6.1). This will help them understand how various types of information are interrelated and contribute to the decision making process.

The appointing party must articulate the information requirements transparently to all associated organisations and individuals. These requirements are typically detailed in four essential documents: organisational information requirements (OIR), asset information requirements (AIR), project information requirements (PIR), and exchange information requirements (EIR). The definitions provided on this section are based on ISO 19650-1.

Organisational information requirements (OIR)

These highlight the information needed to meet strategic objectives within the appointing party's organisation. This includes such elements as strategic business operations, asset management, portfolio planning, regulatory responsibilities and policy making. It is essential to note that OIR serve a strategic role in the business and are not tied to any particular project or appointment.

Asset information requirements (AIR)

The asset information requirements (AIR) outline the managerial, commercial and technical aspects necessary for asset information production. These requirements guide the delivery team's methods and are sufficiently detailed to inform asset management decisions. The AIR are prepared for each trigger event in asset operation and may include security considerations. In supply chains, lead appointed parties can further divide or augment AIR for their specific needs. Across various management appointments, all AIR should form a unified set of requirements, sufficient for addressing asset-related organisational information requirements (OIR).

Figure 6.1 Hierarchy of information requirements (adapted from BSI, 2019; Icons: M.Style/Shutterstock)

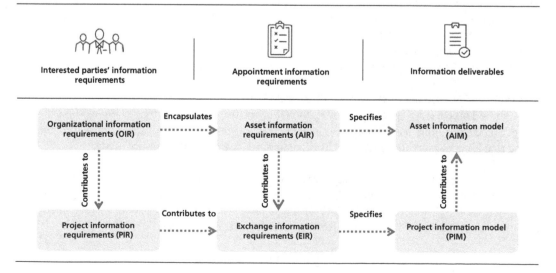

Project information requirements (PIR)

This specify the information required to meet the strategic goals set by the appointing party for specific built asset projects. Information is sourced from both project and asset management processes. For each key decision point in the project, a tailored set of information requirements should be created. Where clients are involved in several projects, a standard set of PIR can be developed and adapted as needed.

Exchange information requirements (EIR)

The EIR detail managerial, commercial and technical guidelines for project information. They include the information standard, the production methods and procedures to be implemented by the delivery team and the information needed to meet the PIR.

The EIR should correspond to key project milestones and should be incorporated in all relevant appointments. They are adaptable and can be subdivided or augmented by different parties along the supply chain.

Across different appointments in a project, the EIR should form a unified, comprehensive set of guidelines to sufficiently meet the project information requirements (PIR).

Asset information model (AIM)

This supports the strategic and daily asset management processes of the appointing party. It can offer valuable information at the start of the project delivery process, including equipment registers, total maintenance costs, installation and maintenance dates, property ownership details and any other relevant information that needs systematic management.

Project information model (PIM)

The PIM is essential for both the project's execution and long-term asset management. It serves as a comprehensive archive, containing key details, such as project geometry, equipment locations, performance criteria, construction methods, schedules, costs and maintenance needs for installed systems and components.

6.4.   Appointing party responsibilities

The appointing parties must clearly state their information needs in project documents. This ensures that the delivery team delivers the information that the appointing party needs and helps to avoid any unnecessary costs. Requesting information that isn't going to be used or that doesn't contribute to the maintenance and performance of the project is a waste of resources. Therefore, appointing parties should carefully consider their information needs, focusing on details that will benefit the end user and help with the effective management and maintenance of the assets.

The ISO 19650-2 standard (BSI, 2021) offers specific guidelines for the actions that the appointing party should take before appointing any individuals, as well as actions to be completed during the project. The appointing party should follow these guidelines to ensure readiness for the project and check that all necessary information is available before any appointments are made.

This section does not replace the need to read ISO 19650-2 to fully understand the details of each clause. Also, for a detailed understanding of these clauses and their application to specific project roles, consult the resources from the UK BIM Framework (2021). They provide relevant insights and examples related to ISO 19650-2 for various roles within a project.

A summary checklist of the appointing party responsibilities is given in Table 6.8.

Table 6.8 Checklist: appointing party responsibilities

Assessment and need (Clause 5.1)
☑ Appoint individuals for information management.
☑ Establish project information requirements (PIR).
☑ Set information delivery milestones.
☑ Integrate organisation-specific information standards.
☑ Determine methods and procedures for information production.
☑ Identify reference information and shared resources.
☑ Establish common data environment (CDE).
☑ Prepare project's information protocol.

Invitation to tender (Clause 5.2)
☑ Specify exchange information requirements (EIR).
☑ Gather reference information or shared resources.
☑ Set tender response requirements and evaluation criteria.
☑ Compile invitation to tender information.

Appointment (Clause 5.4)
☑ Confirm delivery team's BIM Execution Plan: lead appointed party (with awareness of appointing party).
☑ Establish master information delivery plan (MIDP): lead appointed party (with awareness of appointing party).
☑ Complete lead appointed party's appointment documents.

Mobilisation (Clause 5.5)
☑ Mobilise information technology: lead appointed party (with awareness of appointing party).

Information model delivery (Clause 5.7)
☑ Review and accept the information model.

Project close-out (Clause 5.8)
☑ Archive the project information model.
☑ Capture and document lessons learned for future projects.

Assessment and need (Clause 5.1)
Appoint individuals to undertake the information management function (Clause 5.1.1)
OBJECTIVE
To appoint individuals or parties for managing project information effectively throughout the project.

DESCRIPTION
The appointing party is responsible for effective information management throughout the project. For this purpose, specific individuals within the organisation can be assigned to manage information or a prospective lead appointed party or third party can be designated to handle this role. If opting for the latter, a scope of services should be clearly established.

TASK OWNER
The appointing party.

ACTIONS

- *Internal assignment.* Assign specific individuals within the organisation.
- *External designation.* Designate a prospective lead appointed party or third party if internal resources are unavailable.

Establish the project's information requirements (Clause 5.1.2)

OBJECTIVE

To ensure that the appointing party has the necessary information to make informed decisions at key points throughout the project.

DESCRIPTION

The appointing party establishes project information requirements (PIR) to address questions that need answers at key decision points. Note that the PIR inform the exchange information requirements (EIR).

TASK OWNER

The appointing party.

ACTIONS

- Define Project Information Requirements (PIR) mindful of scope, use, work plan, procurement, decision points, and appointing party's questions for informed decisions.
- Ensure that the PIR informs the exchange information requirements (EIR).

Establish the project's information delivery milestones (Clause 5.1.3)

OBJECTIVE

To set clear milestones for the delivery of project information.

DESCRIPTION

The appointing party establishes project milestones for exchanging information, in accordance with the project's plan of work.

TASK OWNER

The appointing party.

ACTIONS

- Set project milestones for information exchange.
- Align milestones with the project's plan of work.

Establish the project's information standard (Clause 5.1.4)

OBJECTIVE

To establish a standardised framework for information exchange and usage.

DESCRIPTION

The appointing party needs to integrate organisation-specific information standards in the project's information standard.

TASK OWNER

The appointing party.

ACTIONS
- Integrate organisation-specific standards.
- Develop a unified project information standard.

Establish the project's information production methods and procedures (Clause 5.1.5)

OBJECTIVE

To ensure that the information meets the organisation's requirements, addressing all aspects of the project.

DESCRIPTION

The appointing party is responsible for setting any specific methods and procedures for the production of information, based on the organisation's needs.

TASK OWNER

The appointing party.

ACTIONS
- Set methods and procedures for information production.
- Align with organisational needs.

Establish the project's reference information and shared resources (Clause 5.1.6)

OBJECTIVE

To ensure that the prospective lead appointed party has access to necessary reference materials and resources.

DESCRIPTION

The appointing party should identify reference information and shared resources to be shared with the prospective lead appointed party during the tender process.

TASK OWNER

The appointing party.

ACTIONS
- Identify reference information for the tender process.
- Share resources with the prospective lead appointed party.

Establish the project's common data environment (Clause 5.1.7)

OBJECTIVE

To facilitate secure and efficient information sharing.

DESCRIPTION

The appointing party must establish a common data environment (CDE) for the project to ensure a collaborative exchange of information. The CDE should be ready before the tender information is released, to facilitate secure information sharing.

TASK OWNER

The appointing party.

ACTIONS

- Establish a common data environment (CDE).
- Ensure that the CDE is ready before the tender process.

The appointing party can either delegate the management of the project's CDE to a third party early on or transfer it to an appointed party later. In either scenario, it's advised that the appointing party outlines both functional and non-functional requirements for the CDE.

Establish the project's information protocol (Clause 5.1.8)

OBJECTIVE

To manage and protect information across all appointments.

DESCRIPTION

The appointing party must prepare the project's information protocol, which will be incorporated in all appointments.

TASK OWNER

The appointing party.

ACTIONS

- Prepare the project's information protocol.
- Incorporate the protocol in all appointments.

An information protocol that can be used is available in the resources of the UK BIM Framework (2023) website.

Invitation to tender (Clause 5.2)

Establish the appointing party's exchange information requirements (Clause 5.2.1)

OBJECTIVE

To clearly define information exchange expectations.

DESCRIPTION

The appointing party is responsible for specifying the exchange information requirements that the prospective lead appointed party is expected to fulfil during the appointment.

TASK OWNER

The appointing party.

ACTIONS

- Specify exchange information requirements for the appointments.
- Align with organisational and project needs.

Note that the exchange information requirements (EIR) are based on the appointment rather than the project, as the lead appointed party can expand the EIR when appointing the task teams.

Assemble reference information and shared resources (Clause 5.2.2)

OBJECTIVE

To provide essential reference information and resources to the lead appointed party.

DESCRIPTION

The appointing party should gather the reference information or shared resources to be provided to the prospective lead appointed party during the tendering process and appointment.

TASK OWNER

The appointing party.

ACTIONS

■ Gather reference information and shared resources.

■ Provide these to the lead appointed party during tendering.

The information should be provided through the common data environment and the suitability of each piece of information should be identified using appropriate metadata to ensure correct use of the information.

Establish tender response requirements and evaluation criteria (Clause 5.2.3)

OBJECTIVE

To establish clear criteria for evaluating tender responses.

DESCRIPTION

The appointing party sets the minimum requirements that the prospective lead appointed party must meet in the tender response.

TASK OWNER

The appointing party.

ACTIONS

■ Set minimum requirements for tender responses.

■ Define evaluation criteria for assessing tenders.

Compile invitation to tender information (Clause 5.2.4)

OBJECTIVE

To ensure a comprehensive and informative tender process.

DESCRIPTION

The appointing party must assemble the relevant information to be included in the tender package.

TASK OWNER

The appointing party.

ACTIONS

■ Assemble information for the tender package.

■ Include exchange information requirements, reference information, response criteria and other documents as highlighted in the ISO 19650-2.

Appointment (Clause 5.4)

Confirm the delivery team's BIM Execution Plan (Clause 5.4.1)

OBJECTIVE

To ensure that the delivery team's plan meets the project requirements.

DESCRIPTION

The primary responsibility for this clause falls to the lead appointed party, but it's essential for the appointing party to be aware of it. The appointing party should be kept informed about the delivery team's BIM Execution Plan and check that it meets any specific requirements.

TASK OWNER

The lead appointed party.

ACTIONS

- Review the delivery team's BIM Execution Plan.
- Confirm alignment with project requirements.

Establish the master information delivery plan (Clause 5.4.5)

OBJECTIVE

To ensure that the MIDP meets the appointing party's requirements.

DESCRIPTION

Although the lead appointed party primarily holds responsibility for this clause, it is crucial for the appointing party to be well-informed about the content of the master information delivery plan (MIDP) and check that it meets any specific requirements.

TASK OWNER

The lead appointed party.

ACTIONS

- Ensure the creation of a master information delivery plan (MIDP).
- Confirm that the MIDP meets project requirements.

Complete the lead appointed party's appointment documents (Clause 5.4.6)

OBJECTIVE

To facilitate effective management and compliance during the appointment.

DESCRIPTION

The appointing party needs to ensure that the appropriate documents are included in the appointment for the lead appointed party and managed by change control during the entire duration of the appointment.

RESPONSIBILITY

The appointing party.

ACTIONS

- Include necessary documents in the lead appointed party's appointment.
- Manage documents using change control throughout the appointment.

In preparing the appointment documents, the appointing party is responsible for reviewing and understanding the approach of the lead appointed party in delivering on these requirements, as well as for maintaining ongoing collaboration and vigilance regarding compliance throughout the appointment.

Mobilisation (Clause 5.5)
Mobilise information technology (Clause 5.5.2)
OBJECTIVE

To ensure effective use of information technology in project mobilisation.

DESCRIPTION

This is another example that illustrates that although the lead appointed party is responsible for this clause, the appointing party should also understand how information technology is structured in the team's mobilisation plan and should also have the necessary training and access to relevant platforms.

TASK OWNER

The lead appointed party.

ACTIONS

▪ Understand the structure of information technology in the team's mobilisation plan.

▪ Ensure training and access to relevant platforms.

Information model delivery (Clause 5.7)
Review and accept the information model (Clause 5.7.4)
OBJECTIVE

To ensure information compliance and accuracy.

DESCRIPTION

The appointing party should review the information provided in accordance with the project's methods and procedures. The appointing party will then either accept or reject the information, and instruct the lead appointed party to make any required amendments before resubmitting.

TASK OWNER

The appointing party.

ACTIONS

▪ Review the information model.

▪ Accept or request amendments as needed.

Project close-out (Clause 5.8)
Archive the project information model (Clause 5.8.1)
OBJECTIVE

To preserve project information for future use and reference.

DESCRIPTION

On acceptance of the final project information model, the appointing party is tasked with archiving the information within the common data environment, following the project's information production methods and procedures.

TASK OWNER

The appointing party.

ACTIONS

- Archive the final project information model to ensure long-term accessibility and reference.
- Adhere strictly to the project's information production methods to maintain consistency and quality.

Capture lessons learned for future projects (Clause 5.8.2)

OBJECTIVE

To improve future project processes and outcomes.

DESCRIPTION

The appointing party, in collaboration with the lead appointed party, must document lessons learned during each stage of the project and store them in a knowledge repository for future reference.

TASK OWNER

The appointing party.

ACTIONS

- Document lessons learned during the project.
- Store lessons in a knowledge repository for future reference.

6.5. Lead appointed party: tender response and appointment
Content of the prospective lead appointed party tender response

The primary goal of this section is to provide insights and clarify the specific documentation and approaches that need to be taken into account when a prospective lead appointed party submits relevant information as part of the tender stage.

In previous observations, I have identified that it's fundamental to have a detailed understanding of the expectations set by the appointing party. These expectations are documented during the tender stage, specifically as outlined in Clause 5.2.4 of ISO 19650-2, titled 'Compile invitation to tender information', which falls to the responsibility of the appointing party. Additionally, it is imperative that the prospective lead appointed party actively seek any required clarification to prevent any potential misunderstandings. By achieving such clarity, the prospective lead appointed party will be able to deliver on these expectations with precision and efficiency.

This precise understanding will enable the lead appointed party to accurately communicate the project's BIM requirements to the different task teams involved during the tender stage. It will assist in selecting the right task teams for the job and prevent any potential extra charges incurred from task teams who do not support the project's culture and capabilities.

To support the communication of the BIM approach to the appointing party and requirements to the different task teams during the tender stage, the prospective lead appointed party must prepare the pre-appointment BIM Execution Plan. This document should respond to the appointing party's exchange information requirements (EIR) and any relevant BIM documentation supplied during the tender stage.

If the appointing party hasn't supplied any documents indicating a need for BIM or expressed an interest in implementing it, there's no requirement to submit a pre-appointment BIM Execution Plan or any other documentation discussed in this section. However, if the lead appointed party has internal policies to implement BIM, regardless of the appointing party's preferences, it's essential to follow the appropriate process and content outlined in this section. This will clearly communicate BIM requirements to the task teams tendering for the project and help mitigate risks during the project.

Additionally, it could be beneficial to express your intention to the appointing party of implementing BIM in the project, highlighting the best practices and potential advantages and explaining how BIM will aid in project delivery, while also outlining the specific data you aim to collect to support your policy strategy, which can be valuable for the appointing party.

Compile the delivery team's tender response (Clause 5.3.7)

The ISO 19650-2 standard (BSI, 2021) specifies the documents the prospective lead appointed party should submit for inclusion within the delivery team's tender response:

- The pre-appointment BIM Execution Plan (Clause 5.3.2).
- The capability and capacity summary (Clause 5.3.4).
- The proposed mobilisation plan (Clause 5.3.5).
- The risk register (Clause 5.3.6).

Next, I outline the contents of each document that the prospective lead appointed party needs to prepare as part of the tender submission.

Establish the delivery team's (pre-appointment) BIM Execution Plan (Clause 5.3.2)

As we have seen, and according to the information management process defined in ISO 19650-2, the prospective lead appointed party is responsible for preparing the delivery team's pre-appointment BIM Execution Plan, to be included in the candidate's response to the tender.

During the tender stage, the pre-appointment BIM Execution Plan provides a clear and comprehensive outline of how the prospective lead appointed party and the task teams intend to implement the BIM process throughout the various phases of the project. It showcases the proposed approach, capability, capacity and expertise of the prospective lead appointed party and the task teams, to support the expectations and requirements of the appointing party.

To develop the pre-appointment BIM Execution Plan, the prospective lead appointed party must collaborate with the known task teams involved at the tender stage. This teamwork ensures a thorough and accurate plan, outlining the implementation of the BIM process throughout the project's lifecycle.

All delivery team members must understand the client's expectations and requirements, and the specific details of BIM implementation, before committing to any work. The prospective lead appointed party and task teams should avoid committing to unclear aspects, as this can result in misunderstandings, delays and additional costs in the future. Therefore, as part of the tender stage, it is crucial for the prospective lead appointed party to take the necessary steps to ensure a clear understanding of the BIM expectations for the project. To accomplish this, it is necessary to raise the necessary tender queries to clarify any aspects of the BIM requirements and ensure a full understanding of the client's BIM expectations.

It is important to note that, on occasion, the appointing party's responses to these tender queries might not be provided in a timely manner or might not fully address the concerns and questions raised. In such cases, it is the responsibility of the prospective lead appointed party to incorporate any comments and concerns in the pre-appointment BIM Execution Plan and the delivery team's risk register. These comments will inform the appointing party of any required modifications or clarifications needed after the appointment to ensure that the relevant parties can produce and deliver the information efficiently and securely.

In developing the pre-appointment BIM Execution Plan, the prospective lead appointed party should consider the following factors, in line with ISO 19650-2.

- The project directory will include the names of the individuals who will be responsible for undertaking the information management function within each task team. Additionally, I recommend to incorporate the contact information of the project leaders for each task team as part of the pre-appointment BIM Execution Plan.

- The information delivery strategy for the delivery team includes several key components. These include the team's approach to meeting the appointing party's exchange information requirements, assigning responsibilities following the information management assignment matrix, a set of objectives or goals for collaborative information production, an overview of the team's organisational structure.

- The delivery team will outline the proposed federation strategy for bringing together information from various sources and maintaining its accuracy and consistency throughout the project. They will need to take into account the fact that various task teams might use different software for creating content and figure out ways to resolve any related problems.

- Describe the high-level responsibility matrix that assigns each component of the information to a specific team. List the key outputs that are associated with each component, so that all stakeholders know who is accountable for delivering certain pieces of information, and what they are expected to produce as the design model evolves at each stage. This matrix should correspond to the scope of work agreed with each stakeholder.

- Propose changes to the project's information production methods to effectively capture existing asset information, generate, review, approve and authorise information, ensure information security and distribution, and deliver information to the appointing party.

- Propose additions or amendments to the project's information standard required by the delivery team to facilitate effective exchange of information between task teams, distribution of information to external parties and delivery of information to the appointing party.

- It is also important to specify the proposed software, hardware and IT infrastructure that the delivery team plans to utilise throughout the project. This is necessary for showing that the delivery team has the necessary technology and resources to efficiently manage the project's information. I encourage the use of tools that everyone can access and navigate around the federated model without needing special licences, hardware or extensive training. The goal is to make the information readily available to everyone on the team.

Ultimately, the primary objective of the pre-appointment BIM Execution Plan is to clearly communicate to the appointing party the team's capabilities in meeting the project requirements and expectations. It serves to eliminate potential misunderstandings that might arise during the delivery of the project after the appointment.

After the completion of the tender stage, it is crucial to allocate time to gather feedback from the appointing party. This feedback holds significant value as it offers insights into areas where the pre-appointment BIM Execution Plan can be enhanced.

Additionally, knowing what worked well, and what didn't, can provide your business with a competitive advantage in future tenders. By understanding your strengths and weaknesses, you can enhance your responses and position your business more effectively to secure contracts. Therefore, having a data analytics strategy in place to capture this feedback is crucial. By analysing the data, you can identify trends and patterns that inform your business decisions, enabling you to make better choices in tender responses.

Furthermore, capturing feedback demonstrates to appointing parties that you value their opinions and are committed to continuous improvement. This fosters trust and strengthens your relationship with each appointing party, which can be advantageous for future business opportunities.

It is important to keep in mind that the capability and capacity assessment summary, mobilisation plan and information delivery risk assessment should be completed during the tender stage.

The different clauses that are part of the prospective lead appointed party's tender response are typically presented as separate documents. However, to minimise the amount of documentation and decrease the likelihood of teams not reading separate documents, I prefer to consolidate information. I like to include the capability and capacity assessment summary, mobilisation plan and BIM-related information delivery risk assessment within the pre-appointment BIM Execution Plan (BEP). I find that this makes it easier for teams to access information without having to review separate documents, and minimises the risk of misplacing information.

It is important to note that the overall project risk register, however, will be a separate document and will also encompass additional risks not related to the BIM process. This will involve input from different team members within the lead appointed party and the selected task teams. Therefore, any BIM-related information delivery risk assessment needs to be incorporated in the overall risk register. I understand that this might initially result in duplication that needs to be updated and managed throughout the project, but this is my personal preference as it facilitates communication of the BIM-related information delivery risk during the project.

Assess and establish the delivery team's capability and capacity (Clauses 5.3.3 and 5.3.4)
As described in clause 5.3.3, during the tender stage, each task team must complete a capability and capacity assessment based on the appointing party's exchange information requirements and the proposed pre-appointment BIM Execution Plan.

The prospective lead appointed party will review and consider each assessment completed by the selected task teams. Gaining insight into the task teams' experience and expertise in handling project information is vital for selecting the most suitable teams and ensuring a smooth project delivery.

A summary of the teams' capabilities and any required training, if necessary, should be included as part of the pre-appointment BEP issued to the appointing party as a component of the tender response, in accordance with clause 5.3.4. By integrating the capability and capacity summary of the task teams in the pre-appointment BIM Execution Plan, the delivery team gains visibility and a comprehensive understanding of each task team's skills and capacity, and the support required for successful project delivery.

A lengthy assessment might not necessarily provide a comprehensive understanding of the capabilities and capacities of the task team. In fact, the longer the assessment, the more likely it is to encounter resistance in completion, and the less likely it is to be completed accurately during the tender stage. Therefore, I discourage the use of these lengthy assessments that have been prevalent in the industry for years and instead suggest focusing the assessment on relevant questions.

I also believe it would be more insightful to have a conversation with the task team rather than relying solely on questionnaire responses. It is necessary to request and review evidence that corroborates their responses during these discussions. Additionally, as previously discussed, it is highly valuable to evaluate feedback on the task team's performance in previous projects before making any assignments.

The assessment should encompass the following three areas of the business, according to Clause 5.3.3 of ISO 19650-2.

TASK TEAM'S INFORMATION MANAGEMENT CAPABILITY AND CAPACITY
- Number of team members with relevant information management experience, needed to meet the project's delivery strategy.
- Education and training available to task team members.

TASK TEAM'S INFORMATION PRODUCTION CAPABILITY AND CAPACITY
- Number of team members with relevant information production experience, needed to meet the project's information production methods and procedures.
- Education and training available to task team members.

TASK TEAM'S INFORMATION TECHNOLOGY (IT) AVAILABILITY
- Suggested IT solutions, their purpose, and software versions.
- Security certifications and measures in place to manage information securely.

Establish the delivery team's mobilisation plan (Clause 5.3.5)
The mobilisation plan should be included in the tender response and should cover all the necessary activities required to set up the project for success from the outset.

The mobilisation plan is not just a one-time document; it is a living document that should be updated throughout the project. This means registering the completion of any of the activities listed as part of the mobilisation plan, as well as recording any additional comments that can provide extra information about the tasks completed. This will help to ensure that the plan remains relevant and up to date and that all stakeholders are kept informed of the project's progress.

The following sections are based on the points included in ISO 19650-2, Clause 5.3.5.

- *Information production methods and procedures.* Test and document the proposed methods and procedures; this includes capturing existing asset information, generating, reviewing or approving new information, ensuring information security and distribution and delivering information to the appointing party.

- *Information exchanges.* To ensure effective information exchange in the project, the mobilisation plan includes a trial period for the project team to test information exchanges between task teams and to test information delivery to the appointing party.

- *Common data environment.* To promote efficient collaboration, the project's common data environment (CDE) will be configured and tested. A clear folder structure, metadata and security measures will be established, and the project team will be invited to participate, with workflows agreed to clarify responsibilities. Additionally, requests for information (RFIs) and design development will be defined during this stage.

- *Information technology.* The mobilisation plan includes the procurement, implementation, configuration and testing of additional software, hardware and IT infrastructure, including tools such as model viewers, site imagery capture systems, and other solutions, to support the project. This involves configuring them to meet project needs and thoroughly testing them to ensure correct functioning and the interoperability of different file formats.

- *Shared resources.* Develop shared resources such as a point cloud survey for gathering accurate data, documentation and templates like the TIDP or custom libraries. These resources ensure that all team members have equal access to the necessary information, streamlining the project's workflow and promoting efficient teamwork.

- *Training and support.* It is crucial to include in the plan the provision of suitable training for both internal and external members of the project delivery team, with a focus on process and technical skills. Also, you need to evaluate whether additional team members need to be recruited to meet the necessary capacity and to assist organisations that join the project in fulfilling its requirements.

Establish the delivery team's risk register (Clause 5.3.6)

As part of the tender response, the lead appointed party will identify potential risks that can affect the delivery of the information according to the appointing party's requirements and the particularities of the project.

The following points are based on ISO 19650-2, Clause 5.3.6. However, it is important to note that the specific list may vary depending on the project requirements.

- *Assumptions made regarding the exchange information requirements by the delivery team during the tender stage.* Examples include unclear asset data requirements and asset tag responsibilities. This can lead to misaligned expectations, potential delays and issues in obtaining data from different task teams if communication is not clear during the appointment.

- *Adherence to the appointing party's timelines for the delivery of project information.* For example, the delivery team might struggle to meet tight deadlines, owing to the novated design team's lack of capacity or capability to deliver project requirements. Other risks might include delays in conducting surveys, assessing the impact on existing buildings or dealing with asbestos or slab deflections. There might also be risks associated with security restrictions on managing project information.

- *The details outlined in the project's information protocol.* For example, the delivery team might not fully agree with the protocol, owing to gaps or contradictions with the project scope.

- *Complying with the proposed information delivery strategy.* For example, the proposed common data environment for sharing information might lack the necessary workflows for approving information or might not comply with the project's security requirements, such as the requirement to keep servers within the UK.

- *Implementing the project's information standard and information production methods and procedures.* For example, the delivery team might face challenges when transitioning from their existing processes to the appointing party's requirements.

- *Amendments to the project's information standard.* For example, the delivery team might suggest the use of an open file format for information exchange. However, if the appointing party is reluctant to accept the change from the specified file format, this could lead to communication breakdowns and delays in information sharing.

- *Mobilisation of the delivery team to meet the required capability and capacity.* For example, if a key team member with specialised BIM expertise leaves the project unexpectedly, the task team might struggle to maintain the necessary level of capability and capacity, potentially leading to project delays or quality issues.

Documentation for the lead appointed party appointment

The purpose of this section is to aid in understanding the information that should be smoothly incorporated in the formal agreement between the appointing party and the lead appointed party, following a successful tender. This information is crucial for determining the appointment and fostering a productive working relationship between the two parties.

The lead appointed party must always stay vigilant, ensuring that the document contains the latest information. In the fast-paced environment of tendering, details can often change quickly. Therefore, any modifications, additions or omissions in the appointing party's exchange information requirements are promptly communicated and acted on. This communication process isn't simply about acknowledgement; it involves taking these changes on board, critically evaluating them and integrating them where appropriate.

Complete the lead appointed party's appointment documents (Clause 5.4.6)

The documents required for inclusion in the lead appointed party's appointment are:

- The BIM Execution Plan (Clause 5.4.1).
- The MIDP (Clause 5.4.5).
- The exchange information requirements (Clause 5.2.1).
- The project's information standard, including any amendments (Clause 5.1.4).
- The project's information protocol, including any amendments (Clause 5.1.8).

Although it is not included in Clause 5.4.6, some may consider it important to include in the appointment the reference information and shared resources, as well as the methods and procedures, including any amendments.

The inclusion of this documentation provides the backbone for the appointment document and the resulting responsibilities of the lead appointed party. Every single facet must be diligently reviewed, accurately integrated and subsequently maintained using change control.

Next, I outline the contents of each document that the prospective lead appointed party needs to prepare as part of the appointment phase.

Confirm the delivery team's BIM Execution Plan (Clause 5.4.1)

After the tender stage, the successful lead appointed party is responsible for completing the BIM Execution Plan. This involves working collaboratively with all parties involved in the project to ensure that the plan is comprehensive and meets the requirements of the appointing party.

As part of this process, the pre-appointment BIM Execution Plan must be updated to provide additional information based on any clarifications provided by the appointing party. This ensures that the plan accurately reflects the needs and expectations of the appointing party and helps to minimise any misunderstandings or delays that could arise during the project.

The following should be considered when preparing the BIM Execution Plan, in line with ISO 19650-2.

- It is necessary to confirm the names of individuals who will be responsible for information management within each task team and capture the project lead for each task team. This helps identify the roles and responsibilities of the team members and ensures accountability for the management of project information. The project directory can be used by other teams joining the project to quickly find the correct points of contact.

- If any changes have been identified from the pre-appointment BEP, the delivery team's information delivery strategy should be updated.

- If any changes have been identified from the pre-appointment BEP, the high-level responsibility matrix should be updated.

- It is necessary to confirm and document the delivery team's proposed information production methods and procedures.

- It is important to reach an agreement with the appointing party on any additions or amendments to the project's information standard.

- The schedule of software, hardware and IT infrastructure that the delivery team will use during project delivery should be confirmed, and it should be clear what tools will be used by each task team.

When creating a BIM Execution Plan, my suggestion is to prioritise simplicity. It is important to minimise the number of separate documents and use plain language. The plan should cover all relevant points, provide clarity on expectations and requirements and address potential concerns. Additionally, it should be updated with new information as the project progresses to remain effective. The plan should also be easily accessible to readers with varying levels of expertise. The goal is to have people embrace the plan rather than feel intimidated or overwhelmed by it. To achieve this, it is important to use simple terms that are easy to understand, even for those who are not BIM specialists. By emphasising clarity and accessibility in the plan, we can increase the likelihood of successful implementation and utilisation by all members of the delivery team.

Establish the delivery team's detailed responsibility matrix (Clause 5.4.2)
The lead appointed party is responsible for establishing a detailed responsibility matrix. This matrix identifies:

- The information to be produced.
- The stage when the information will be exchanged.
- The parties involved in the exchange.
- The task team responsible for its production.

By defining the required information, the delivery team can work towards the same goal, preventing unnecessary duplication of information.

In my experience, when developing the detailed responsibility matrix, the lead appointed party should consider such factors as:

- The task information delivery plan.
- The exchange information requirements.
- Consider the capabilities of the task team to deliver the requirements at each stage.
- Any dependencies in the information production process.

A well-defined responsibility matrix ensures that the delivery team knows whom to approach when questions arise about a particular aspect of the project. It also provides clarity on who is responsible for producing the required information at each stage, thereby avoiding conflicts during the project. Table 6.9 is an example of a detailed responsibility matrix. However, this is just an example, and there are various ways to present the necessary content.

Establish the master information delivery plan (Clause 5.4.5)
The lead appointed party is responsible for compiling the task information delivery plan (TIDP) from each task team and ensuring that it is kept up to date to form the delivery team's master information delivery plan (MIDP).

In doing this, the following factors should be considered:
- The detailed responsibility matrix.
- The project programme and dependencies between task teams.
- The time required for the lead appointed party to review and authorise the information.
- The time that the appointing party will need to review and accept the information.

After establishing the MIDP, the lead appointed party is responsible for:
- Ensure the MIDP remains updated throughout the different stages of the project.
- Incorporate the new TIDP from new task teams involved in the project.
- Understand the impact on the MIDP if some information is delivered late.
- Use tools to support visualising and measuring the impact of changes and the progress made.
- Notify the appointing party of any potential issues that may impact the project's information delivery milestones.

The other documents, which are produced by the appointing party and have been discussed in Section 6.4, need to be incorporated in the appointment, along with any agreed additions or amendment, as follows.
- Establish the appointing party's exchange information requirements (Clause 5.2.1).
- Establish the project's information standard (Clause 5.1.4).
- Establish the project's information protocol (Clause 5.1.8).

Table 6.10 is a checklist for the prospective lead appointed party. It specifies which documents need to be included in the tender response, as well as which documents will be required for the appointment if the tender is successful.

Table 6.9 Detailed responsibility matrix

Royal Institution of Chartered Surveyors: New rules of measurement		Spatial coordination — Stage 3 Author	Input	Technical design — Stage 4 Author	Input	Manufacturing and construction — Stage 5 Author	Input
1.1 Substructures	1.1.1 Standard foundations	Consultant A	n/a	Consultant A	n/a	Consultant A	n/a
	1.1.2 Specialist foundation systems	Consultant A	n/a	Consultant A	n/a	Consultant A	n/a
	1.1.3 Lowest floor construction	Consultant A	n/a	Consultant A	n/a	Consultant A	n/a
	1.1.4 Basement excavation	Consultant A	n/a	Consultant A	n/a	Consultant A	n/a
	1.1.5 Basement retaining walls	Consultant A	n/a	Consultant A	n/a	Consultant A	n/a
2.1 Frames	2.1.1 Structural Steel frames	Consultant A	n/a	Consultant A	n/a	Contractor D	n/a
	2.1.2 Space frames / decks	Consultant A	n/a	Consultant A	n/a	Contractor D	n/a
	2.1.5 Timber frames	Consultant A	n/a	Consultant A	n/a	Contractor D	n/a
3.1 Wall finishes	3.1.1 Finishes to walls	Consultant B	n/a	Consultant B	n/a	Contractor B	n/a
3.2 Floor finishes	3.2.1 Finishes to floors	Consultant B	n/a	Consultant B	n/a	Consultant B	n/a
	3.2.2 Raised access floors	Consultant B	Consultant C	Consultant B	Consultant C	Consultant B	Contractor E
3.3 Ceiling finishes	3.3.1 Finishes to ceilings	Consultant B	n/a	Consultant B	n/a	Consultant B	n/a
	3.3.2 False ceilings	Consultant B	Consultant C	Consultant B	Consultant C	Consultant B	Contractor E
	3.3.3 Demountable suspended ceilings	Consultant B	Consultant C	Consultant B	Consultant C	Consultant B	Contractor E
5.4 Water installations	5.4.1 Mains water supply	Consultant C	n/a	Consultant C	n/a	Contractor E	n/a
	5.4.2 Cold water distribution	Consultant C	n/a	Consultant C	n/a	Contractor E	n/a
	5.4.3 Hot water distribution	Consultant C	n/a	Consultant C	n/a	Contractor E	n/a
	5.4.4 Local hot water	Consultant C	n/a	Consultant C	n/a	Contractor E	n/a
	5.4.5 Steam and condensate distribution	Consultant C	n/a	Consultant C	n/a	Contractor E	n/a
5.5 Heat source	5.5.1 Heat source	Consultant C	n/a	Consultant C	n/a	Contractor E	n/a

Table 6.10 Checklist: tender submission and appointment

Compile the delivery team's tender response (Clause 5.3.7)
- ☑ The pre-appointment BIM Execution Plan
- ☑ The capability and capacity summary
- ☑ The proposed mobilisation plan
- ☑ The risk register

Complete lead appointed party's appointment documents (Clause 5.4.6)
- ☑ The BIM Execution plan
- ☑ The MIDP
- ☑ The exchange information requirements
- ☑ The project's information standard, including any amendments
- ☑ The project's information protocol, including any amendments

6.6. Task team: invitation to tender and appointment

This chapter, like the previous one, seeks to clarify the procedures and methods that should be employed when interacting with prospective task teams during the tender process and on their appointment. This applies to design consultants and subcontractors alike.

To ensure a successful project delivery, it is necessary that all task teams involved in the project not only have a clear understanding of their contractual obligations but also have a thorough comprehension of the information available at that point. This dual understanding allows each task team to plan effectively and estimate the scope of the project and its requirements with precision.

Hence, during the tender stage, it is particularly important to hold in-depth discussions, provide exhaustive clarifications and be transparent with each task team to build trust regarding the available information. This trust allows teams to provide accurate cost estimates based on the project's requirements. This step is designed to mitigate potential conflicts or disagreements that could arise during the project's execution phase. If these are overlooked, task teams might try to offset their costs by reducing the quality of their services, which can have a negative impact on the project's implementation.

Effective collaboration and communication between the various task teams and the lead appointed party play a significant role in ensuring smooth progress of the project and achieving its objectives. To accomplish this, during the tender stage, it is essential for task teams to receive accurate and relevant information, allowing them to fully grasp the project requirements, including the BIM expectations. Providing the necessary information tailored to each task team's role is crucial, as teams can better understand their part in the overall project, leading to a more effective and efficient tender process and improved project execution overall.

This strategy helps to prevent overloading task teams with irrelevant information, which could divert their attention from identifying and reviewing crucial details related to their responsibilities. Avoiding the supply of unnecessary information, eliminating duplications and irrelevant details, can minimise confusion, prevent potential negative consequences such as increased costs or hesitancy in participating in the tender process, and contribute to a smoother and more efficient tender process.

Content of the invitation to tender for the appointed party

While the ISO doesn't specify the information to be provided to the prospective appointed party (task team) during the tender stage, the process could mirror that of the appointing party interacting with a prospective lead appointed party.

When engaging with task teams that have design responsibilities, the aim is to provide the relevant documentation to enable accurate tendering for the project.

This approach to documentation is similar to how the appointing party prepares the tender invitation for the prospective lead appointed party. Consider:

- The lead appointed party's exchange information requirements.
- The milestones for delivering project information.
- The project's information production methods and procedure.
- The project's information standard (including any agreed additions or amendments).
- The project's information protocol (including any agreed additions or amendments).
- The project's BIM Execution Plan.
- The relevant reference information and shared resources (within the project's common data environment), including models and survey information, if available.

In addition to these documents, the lead appointed party should provide the following documentation for the prospective task teams to complete and return as part of the tender response:

- The BIM capability and capacity assessment, to be completed and returned by each task team.
- The project's TIDP, to be populated and returned by each task team.

It's important not to overwhelm task teams with irrelevant or excessive information beyond their scope and responsibilities. Before sharing information, review each task team's specific role and duties to ensure the information shared is relevant. This careful approach promotes efficiency and equips the task teams with the necessary information for accurate responses.

Merely sharing information is not enough; active engagement with each task team in the tender process is key. Arrange discussions to address any concerns or queries the teams might have, helping to clear any uncertainties. These discussions ensure that task teams fully grasp the project expectations, leading to more accurate and optimised responses in their quotations. By clarifying any issues and promoting understanding, you lay the foundation for receiving the best possible responses, contributing to the project's overall success.

While I have addressed most of the documents relevant to the tender stage, the following sections outline the lead appointed party's exchange information requirements and the task information delivery plan, both of which are crucial components of the tender stage.

Establish the lead appointed party's exchange information requirements (Clause 5.4.3)

The lead appointed party should define exchange information requirements (EIR) for each task team appointment, mirroring the approach of the appointing party.

The EIR established by the lead appointed party should specify the precise information required from each task team, ensuring a cascade effect across the relevant task teams.

The lead appointed party should undertake the following actions, as outlined in ISO 19650-2, Clause 5.4.3.

- Define each information requirement by considering:
  - ☐ The appointing party's information requirements.
  - ☐ Any additional information requirements set by the lead appointed party.
  - ☐ Ensure that any additional requirements comply with the business policy and do have a real purpose.
- Determine the level of information need to satisfy each information requirement.
- Set acceptance criteria for each information requirement, considering:
  - ☐ The project's standard for information, along with the methods and procedures for producing project information.
  - ☐ Internal quality assurance requirements set by the lead appointed party.
- Set deadlines for each requirement in relation to the project's information delivery milestones, considering:
  - ☐ Time required to complete the approval workflow specific for the project.
- Identify the supporting information that the task teams may need to fully understand or evaluate each information requirement or its acceptance criteria. Consider:
  - ☐ Information provided by the appointing party and information produced before the appointment.
  - ☐ Shared resources and examples of expected deliverables to be used as reference.
  - ☐ References to relevant industry standards.

Establish task information delivery plans (Clause 5.4.4)

Each task team is expected to submit a comprehensive TIDP during the tender stage. This allows the lead appointed party to gain a clear understanding of the information that the task team intends to deliver and the corresponding timelines. By providing this visibility, the lead appointed party can promptly identify and rectify any inaccuracies before finalising the appointments. Additionally, this ensures that all shared information during the project complies with the appropriate naming protocol, thereby avoiding the need to reject and revise information, resulting in smoother project execution.

The TIDP is a fundamental tool to support the information management of the project; therefore, it is important that the lead appointed party consider the following to maximise its benefits:

- Confirming the TIDP contains all the information expected at each stage.
- Notify each task team of the necessary changes to the TIDP.
- Ensure the TIDP is updated during the duration of the project.
- Make shared resources available to avoid delays in information production.
- Ensure that the task team submits the information with the appropriate quality at the agreed milestones.

The TIDP should list and identify, for each information container:

- The name and title of the information container.
- Any predecessors or dependencies.
- The level of information need.

■ The estimated time to produce the information.

■ The responsible information author.

■ Delivery milestones.

As already covered, each task team's TIDP is part of the overall project MIDP, which compiles all the TIDPs within the delivery team. The purpose of the MIDP is to ensure correspondence with the team's schedule and logical sequence of deliverables. Keeping the MIDP up to date is crucial, including any changes in TIDPs or new task teams joining the delivery team. Table 6.11 illustrates a potential approach to defining the content of the TIDP.

In addition to the lead appointed party's exchange information requirements and the task information delivery plan, as previously listed, the following clauses are applicable and should be considered as part of the tender. These clauses, are discussed in detail in Sections 6.4 and 6.5.

■ Establish the project's information delivery milestones (Clause 5.1.3).

■ Establish the project's information standard (Clause 5.1.4).

■ Establish the project's information production methods and procedures (Clause 5.1.5).

■ Establish the project's reference information and shared resources (Clause 5.1.6).

■ Establish the project's information protocol (Clause 5.1.8).

■ Assess task team capability and capacity (Clause 5.3.3).

■ Confirm the delivery team's BIM Execution Plan (Clause 5.4.1).

Content of the appointed party appointment documents

In this instance, ISO 19650-2 (BSI, 2021) provides a clear clause regarding the information that should be incorporated in the task team appointment.

Complete the appointed party's appointment documents (Clause 5.4.7)

It is essential to consider specific information that is tailored to the unique needs of the project and to ensure that all appointment documentation is meticulously prepared. It is important to avoid using templates filled with placeholders like 'to be confirmed' or equivalent, as they can cause confusion and hinder the timely completion of the project.

Table 6.11 Task information delivery plan

Name	Title	Level of information need			RIBA Stage 3: spatial coordination			
		Format	Scale	Content	Author	Predecessor	Delivery milestone	Duration
ABCD-AMA-XX-08-D-A-2300	General arrangement	PDF	1:100	To adequately describe these elements: ■ Fire strategy drawings ■ Accessibility strategy	Michael	Survey information	20-09-2024	20 days

Table 6.12 Checklist: task team tender submission and appointment

Tender invitation to appointed party
☑ Issue the lead appointed party's exchange information requirements and information delivery milestones.
☑ Share the project's information standard, methods and procedure, and protocol, including any amendments.
☑ Provide relevant reference information and shared resources.
☑ Distribute the project's BIM Execution Plan.
☑ Send the BIM assessment template for completion by each task team.
☑ Issue the project's TIDP template for population by each task team.

Complete appointed party's appointment documents (Clause 5.4.7)
☑ Incorporate the lead appointed party's exchange information requirements.
☑ Include the project's information standard and protocol, along with any amendments.
☑ Attach the project's BIM Execution Plan.
☑ Include the agreed TIDP.

The following information should be incorporated when preparing appointments with the relevant task teams and managed through change control for the entirety of the appointment:

- The BIM Execution Plan.
- The TIDP.
- The lead appointed party's exchange information requirements.
- The project's information standard, including any amendments.
- The project's information protocol, including any amendments, ensuring that the content is being updated with the details of each task team.

As per the comment on the lead appointed party's appointment documents, although not included in Clause 5.4.7, some may consider it important to include in the appointment the reference information and shared resources, the methods and procedures, including any amendments, as well as the task team's response to the BIM assessment.

Incorporating these pertinent documents in the contractual obligations of the relevant task teams can help maintain accountability for efficient and effective project delivery. This approach can enhance the overall execution of the project, ensuring that it meets the specified requirements within the allotted time frame and budget.

Table 6.12 is a checklist for the lead appointed party. Its purpose is to assist the estimator and commercial team in ensuring that they provide the necessary documentation during the tender stage to the prospective task team. The table also specifies which documents will be required when appointing the task teams. Before sending these documents at the tender stage or including them in the appointments, make sure they are relevant to each task team.

6.7. Traditional procurement

When you receive information during the tender stage, it is necessary to understand which type of procurement method the project will use – that is, whether it will be traditional or design-and-build. This will define the approach to delivering BIM on the project.

The process that we have seen so far is for a design-and-build project. However, there are several considerations to keep in mind when approaching a traditional procurement method, to avoid issues during the project.

First and foremost, it is essential to comprehend the expectations of the appointing party for the project. If a document of exchange information requirements has been created, outlining the appointing party's BIM expectations, it is crucial to review it during the tender stage, along with all other documentation and information. A BIM Execution Plan should be available, defining the strategy followed by the appointing party's design team to comply with the exchange information requirements.

Additionally, it is essential to ensure that the produced information adheres to and is consistent with the BIM Execution Plan. This step involves reviewing all documentation to ensure compliance with the agreed naming protocol and metadata, as well as the quality of the models and their level of coordination. It is key to ensure that all models produced are accessible and are transferred to the lead appointed party for review during the tender stage.

Having the models at the tender stage, including both the native and Industry Foundation Classes (IFC) schema, not only provides a better understanding of the information quality but also allows other task teams bidding for the works to gain a deeper insight into the project status and work required for the contractor's design portion (CDP) when applicable. Occasionally, the models are not available until after the contract award, disrupting the benefits of BIM and the purpose of collaboration that it aims to achieve.

The data requirements included in the client's exchange information requirements should also be provided and evaluated at the tender stage to ensure that the design team has produced them and that they meet the client's requirements.

Reviewing all this information and having a clear understanding of the appointing party and the work completed by the design team will enable the lead appointed party bidding for the works to have a thorough assessment of the situation and analyse opportunities and risks.

While these actions may seem obvious, in many traditional procurement projects where the client desires to implement BIM on the project, the information produced up to the stage when the lead appointed party becomes involved might not adhere to the client's requirements. There have been scenarios where no BIM Execution Plan had been produced, or even situations when a BIM Execution Plan was available but the information produced by different disciplines did not comply with it.

Therefore, it is important to identify and raise any issues identified in the information available during the tender stage as part of the tender response. The prospective lead appointed party cannot take responsibility for the work completed before the appointment, or, if such a responsibility has been agreed, should at least be aware of it, as rectifying the information produced by the design team to align with the appointing party's expectations might necessitate extra effort and could potentially impact task team appointments. These scenarios exemplify why the appointing party team needs to be involved throughout the project delivery, assessing the work completed by each task team at every stage, ensuring timely rectification and addressing instances of non-compliance.

After reviewing all the information, it is essential to evaluate the contractor's design portion (CDP) elements and ensure that the subcontractors responsible for developing these elements are BIM capable and can provide the required design and data.

Lastly, it is crucial to clarify the responsibility for conducting design coordination with the new CDP and the rest of the design to identify any issues and minimise risks on site. Although the BIM process does not replace the responsibilities to coordinate design that would take place in a traditional 2D project, it is important to clarify the responsibilities at the tender stage to avoid pushbacks during project delivery, as discussed in Section 7.2. This involves working with the lead designer to ensure that everyone understands their responsibilities and can collaborate to deliver a successful project. Table 6.13 is included to assist in reviewing the tender documentation.

Table 6.13 Traditional procurement considerations (continued on next page)

☑ *Verify the appointing party's expectations*. Cross-check the appointing party's exchange information requirements (EIR) document against what's been communicated.

☑ *Examine BIM Execution Plan*. Confirm that the BIM Execution Plan meets the appointing party's exchange information requirements (EIR).

☑ *Document consistency checks*. Perform a meticulous review of all provided documentation to ensure that it complies with the BIM Execution Plan, focusing on verifying its consistency with the appointing party's expectations.

☑ *Assessment of models in both native and IFC schema*. Verify the quality of the models to ensure that they meet the appointing party's requirements and offer insights into the project status and contractor's design portion (CDP).

☑ *Data requirement checks*. Cross-verify the data produced by the design team with the appointing party's stated requirements in the EIR.

☑ *Opportunity and risk assessment*. Check that your understanding of the appointing party's expectations matches the work completed by the design team; identify any disparities as potential opportunities or risks.

☑ *Issue identification and reporting*. Inconsistencies or discrepancies found should be flagged immediately and included in the tender response to the appointing party.

☑ *Contractor's design portion (CDP) verification*. Assess whether the task team responsible for the CDP are BIM capable and whether the team's planned contributions meet the appointing party's requirements.

☑ *Clarification of design coordination*. Verify who will take responsibility for coordinating the new CDP elements with existing design plans, and ensure that this meets the appointing party's expectations.

BIBLIOGRAPHY

BSI (2019) BS EN ISO 19650-1:2018: Organization and digitization of information about buildings and civil engineering works, including building information modelling (BIM). Information management using building information modelling. Part 1: Concepts and principles. BSI, London, UK.

BSI (2021) BS EN ISO 19650-2:2018 & Revised NA: Organization and digitization of information about buildings and civil engineering works, including building information modelling (BIM). Information management using building information modelling. Part 2: Delivery phase of the assets. BSI, London, UK.

NEC (2022). *NEC and CLC Guidance for Dealing with Retention Payments Under NEC3 and NEC4 Contracts.* NEC, London, UK. https://www.constructionleadershipcouncil.co.uk/wp-content/uploads/2022/11/NEC-and-CLC-Guidance-for-Dealing-with-Retention-Payments-Under-NEC3-and-NEC4-Contracts-15.11.22.pdf (accessed 28/11/2023).

UK BIM Framework (2021) ISO 19650 Guidance 2: Delivery Phase. https://ukbimframework-guidance.notion.site/ISO-19650-Guidance-2-Delivery-phase-6124641a84d64bd09d30ad57a506629f (accessed 27/11/2023).

UK BIM Framework (2023) The overarching approach to implementing BIM in the UK. https://www.ukbimframework.org (accessed 24/11/2023).

Amador Caballero
ISBN 978-1-83549-446-2
https://doi.org/10.1680/iceedc.9446207

Chapter 7
Advanced topics and best practices

7.1.   Introduction

In the two previous chapters, I have covered the fundamentals required to understand the BIM process, as well as the roles and responsibilities that are essential for proper information management during the asset delivery phase. The aim of this chapter is to delve into a variety of advanced topics and best practices that are crucial for industry professionals to comprehend and implement. This chapter will explore advanced methods and considerations, introducing emerging trends like the golden thread and digital twins.

One key area that this chapter will address is the important but also controversial topic of clash detection. The aim is to clarify the processes and responsibilities involved, to identify and resolve design conflicts at an early stage. The principle of 'no construction until completed design' will be elucidated to stress the importance of completing the design phase before initiating construction, thereby avoiding ad hoc improvisation or the flawed 'construction first, design later' approach.

Subsequent sections will evaluate the necessity of incorporating digital construction roles in your business and I share some ideas that you may want to consider before seeking BIM certification. The pros and cons of hiring external BIM consultants will also be discussed, offering a balanced perspective to aid informed decision making.

Challenges associated with existing buildings will be examined as well, demonstrating how point cloud surveys can help to address these issues by verifying both the design and the completed works. On a more technical note, I will share some key model requirements that should be considered when a model is to be shared with the team.

I will emphasise the importance of improving the accuracy of operation and maintenance (O&M) documents and handovers in the industry, to better support clients' asset management goals. In alignment with this, I will delve into the asset information model, exploring its reliance on proper information management and its close relationship with the golden thread, both of which are integral to the effective implementation of digital twins.

Overall, the goal is to offer guidance and insights on these advanced topics, providing food for thought as you implement BIM in your projects and business.

7.2.   Clash detection

Utilising 3D models for design coordination

Speaking personally, I am not a fan of the term 'clash detection,' which I prefer to refer to as 'design coordination'. Design coordination and clash detection are essentially similar processes that focus on aligning and ensuring compatibility of all elements within a project's design. Clash detection is commonly associated with the use of 3D models and software to identify potential issues that might not be easily noticeable through traditional 2D coordination methods. The level of design coordination achievable with traditional 2D information is much more limited than can be attained in a 3D environment. However, both clash detection and design coordination share the same objective of enhancing the overall coordination of a design.

The combination of designs from various disciplines in a 3D environment, known as federated models, enables the delivery team to visualise the integration and interaction of different models. This capability is helpful for identifying conflicts, errors and areas for improvement in the design. Moreover, utilising specific software can enhance and expedite the design coordination process by providing a comprehensive and interactive view of a project's design.

Figure 7.1 visualises the large number of different task teams that can contribute to the federated model. The input of all relevant task teams aids in coordinating the design and produces better information for the project.

Overall, the use of 3D models to support design coordination can increase confidence within the delivery team, ensuring proper alignment of all design aspects and early identification and

Figure 7.1 Federated model (Icons: Benvenuto Cellini/Shutterstock)

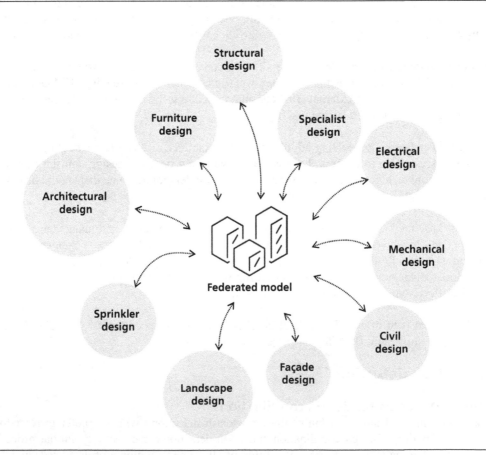

Figure 7.2 Design coordination issue

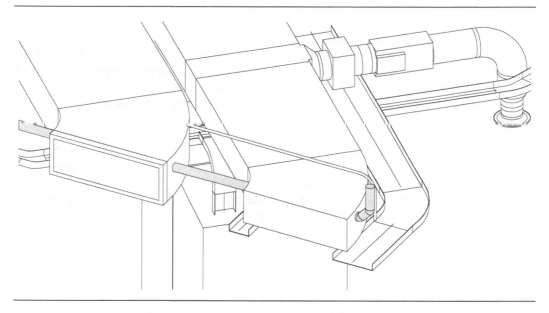

resolution of any issues. This ultimately contributes to more successful and efficient project execution. Figure 7.2 shows an example of a clash that is visible in a 3D environment but might easily be missed in a 2D view. This figure highlights that the pipe and the ductwork run at the same level, necessitating coordination of their positions to avoid issues. On-site resolutions to such problems could impact other trades negatively, which is why prior coordination is crucial.

The role of the lead designer in design coordination

It's common for architects to assume the role of lead designers, taking on the task of conducting design coordination as part of their responsibilities. However, I've come across situations where architects hesitate to lead design coordination when 'clash detection' terminology comes into play. Some might demand additional fees, while others outright refuse to perform this function. Therefore, I tend to avoid using this terminology.

I respect architects and engineers who honestly acknowledge their lack of skills in conducting design coordination using 3D models and specialised software. In such cases, I have been more than willing to provide the necessary support to find a solution that would benefit the project. However, I find it difficult to comprehend why a competent and skilled architect would request an additional fee for performing the design coordination task solely because it involves utilising new technology. This technology, in fact, makes their job easier, more accurate and faster, compared with traditional methods. Design coordination should be done correctly and efficiently, regardless of the methods used. If design errors leading to on-site issues aren't identified, the party responsible for design coordination should be held accountable. So, if technology increases one's productivity and makes the task easier and more efficient, why resist adopting it or increase fees?

Therefore, this subject can indeed be a topic of discussion and negotiation. It's crucial not to presume who will take the lead on design coordination. Instead, you should clarify the responsibilities, and the expectations, including tools and processes to be implemented for the project as part of the BIM Execution Plan before making any appointments, and address any gaps that can lead to undesired design coordination gaps.

The importance of coordination using the models

Design errors can have negative effects on the construction process and the performance of the completed project. These errors can take many forms, including miscalculations, coordination problems, inefficient use of space or the neglect of important requirements. They can compromise the safety, functionality and overall success of the project, highlighting the importance of careful review and collaboration among design professionals.

Incorporating the concept of clash avoidance in your design process can offer substantial benefits. By developing the design in a 3D environment and continuously collaborating with other teams you can proactively identify and address potential clashes before they become actual issues. This pre-emptive approach substantially reduces the number of clashes detected later on, compared with traditional methods that don't involve 3D modelling.

The use of 3D models is a valuable tool for identifying clashes or conflicts that arise between different building components or systems. By leveraging this technology, we can effectively address many design errors, with a particular focus on clashes.

Clashes can be classified as either hard clashes or soft clashes, as illustrated in Figure 7.3. A hard clash occurs when two objects directly collide or occupy the same space. Conversely, a soft clash, which may be less considered and sometimes overlooked, happens when one object encroaches on the designated clearance area required for access or maintenance of another object, even without direct contact. Both types of clash can lead to issues if ignored. Hence, it is crucial to identify and resolve them early in the design process, to ensure a smooth construction phase and optimal performance of the structure.

Strategic issue resolution and resource allocation

The use of specialised software can significantly enhance the identification of clashes and potential non-compliance issues in the design process. By leveraging this technology, teams can efficiently detect clashes based on agreed tolerances, outlined in the BIM Execution Plan. Additionally, the software can be programmed to highlight design elements that do not meet building regulations, ensuring compliance and minimising future complications. For example, the software can be configured to raise an alert if a hallway fails to meet the minimum required width or if a space does not meet

Figure 7.3 Hard and soft clashes

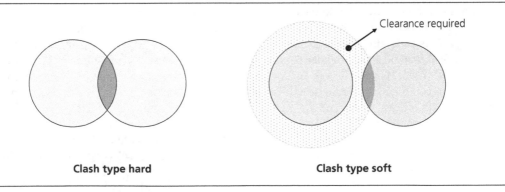

Clearance required

Clash type hard   **Clash type soft**

accessibility standards for individuals with reduced mobility. This automated detection system not only saves time but also improves on accuracy and thoroughness in identifying design issues.

While technology can greatly aid in expediting the design process, it is important to approach it with caution. It's worth noting that not every clash automatically identified by the software is a 'real' clash that requires immediate attention or will directly impact the construction process on site. It is essential for teams to establish the correct rules and parameters within the software to identify and prioritise design issues accurately.

Companies should avoid relying solely on the initial output of the software without verifying its significance and relevance to the actual state of the design. Failure to do so might lead to unnecessary panic and a perception that the design is in a worse condition than it actually is. Proper evaluation and interpretation of the software's results help maintain a balanced perspective and prevent overreaction. As shown in Figure 7.4, a single design error can lead to numerous clashes, identified by the software. Instead of treating each clash as an individual issue, these should be grouped together to indicate that there is only one problem to address. This example demonstrates that the software has flagged dozens of errors, when in reality, there is a single design error related to the height of the services against the timber frame.

Additionally, it is worth noting that it might not always be feasible to resolve all design issues at once. Depending on the complexity of the project and the availability of resources, the team may need to prioritise certain areas or coordinate with external stakeholders to gather necessary information. By understanding these constraints, the team can manage expectations and allocate resources effectively to address design issues in a systematic and strategic manner.

Internal design review and collaboration

Before sharing models for coordination, it is crucial for each design discipline to conduct multiple internal design reviews. These reviews ensure that the design is suitable and well-coordinated with other disciplines' designs. By conducting these reviews, both hard and soft clashes can be addressed, supporting efficient design development. Failing in this diligence can result in complicated coordination efforts and hinder design progress, which can ultimately impact project deadlines.

Once the designs have undergone thorough internal checks and reached a satisfactory level of confidence, they can be shared with the rest of the design team. This sharing takes place at agreed milestones throughout the project, to verify proper integration and ensure that there are no errors or conflicts between the designs of different task teams.

In my experience, identifying design conflicts has been the easier part of the process. The real challenge lies in ensuring that the team effectively resolves and addresses these issues within the agreed deadlines. This is where I have encountered more difficulties, as teams sometimes fail to address the highlighted issues in a timely manner, leading to delays in the coordination process and impacting the project timeline.

To ensure timely resolution of design issues, it is crucial to gain the support and commitment of all task team leaders. They need to understand the importance of taking prompt action to minimise the impact on project deadlines and ensure successful delivery of the project. Strong support from leaders within your organisation is necessary to drive prompt actions from the different task teams. Simply identifying design issues is not enough; it is essential to have the full cooperation and

Figure 7.4 Design error with numerous clashes

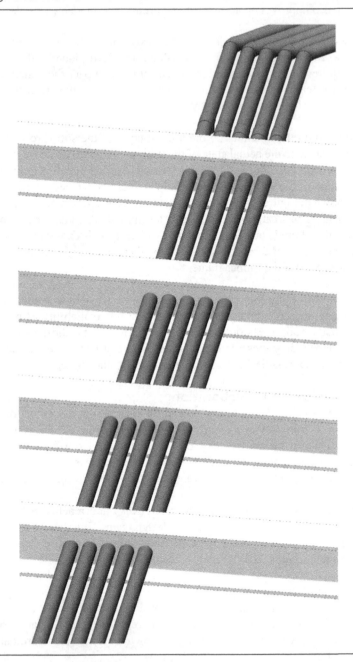

commitment of the teams to address the issues on time and fully reap the benefits of the coordination process.

Coordination processes with digital models

The coordination of the design will be facilitated through the use of digital models, which will be developed and shared at regular intervals throughout the design and construction phase. An example of a 2-week cycle is typically followed, but the schedule can be adjusted to meet the specific requirements of the project, as agreed by the delivery team. Depending on the stage of the project, it may be necessary to conduct checks at specific points in time rather than following a regular schedule for model development and review.

However, the traditional cycle is now evolving, owing to advances in technology. Current technology allows for real-time sharing and communication of issues, enabling prompt resolution as soon as a problem is identified. Furthermore, issues can be easily viewed on model viewers that do not require specialised software or advanced skills to use. This development has been a game-changer in the industry, as it allows team members, even those who are not tech-savvy, to access and review the models and comments effortlessly without needing to have a digital construction specialist in the room.

As a result, the need for dedicated coordination meetings to review all clashes on a screen and discuss solutions for each has been significantly reduced and can become a thing of the past. These traditional coordination meetings can be time-consuming and less engaging for many team members. Instead, only the most critical issues that require input from the entire team will be discussed during the design team meeting (DTM). This approach streamlines the coordination process, saving time and resources for teams with design responsibilities.

Table 7.1 displays the traditional bi-weekly model for the exchange process. It should be noted that this is just one example and that different teams may have alternative approaches, such as uploading all task team models on the same date or requiring additional time for design coordination review. However, it is important to stress that the approach to design coordination is evolving in response to advances in technology. With new technologies, design errors can now be communicated in real time, eliminating the need to wait for a review report before taking corrective action to improve the design.

During the preparation of files for the coordination and model review process, it is crucial to bear in mind several important points. While some of the following examples may not be applicable,

Table 7.1 Information exchange calendar

Mon	Tue	Wed	Thu	Fri	Mon	Tue	Wed	Thu	Fri
Arch Stru			M&E			Design review	DTM		

Arch	Architect design to be uploaded on the common data environment (CDE) and relevant platforms used on the project.
Stru	Structural design to be uploaded on the CDE and relevant platforms used on the project.
M&E	Mechanical and electrical design to be uploaded on the CDE and relevant platforms used on the project.
Design review	Design coordination using the models to be led by the architect as lead designer. Issues to be uploaded on the CDE.
DTM (design team meeting)	Discuss the relevant issues that need team collaboration and agree on the priority actions.

depending on the technology or process that the task team agrees to use and follow, they are worth considering.

- Task teams should each upload their models to the common data environment (CDE), regardless of the amount of work completed since the previous issue, and take accountability to resolve the assigned design issues within the agreed timescales. If no progress has been made, clear communication should be provided to the lead appointed party and the rest of the team, to ensure clarity.

- In the event of major changes to the design model outside of the scheduled exchange intervals, it's crucial to immediately upload the revised model to the CDE with the relevant suitability code. This allows all task teams to incorporate the major change before the regular exchange, preventing significant disparities between models that can impact the coordination process.

- Splash pages or revision schedules should accompany the model upload, clearly indicating any changes made since the previous issue. This helps other task teams to easily identify and review the modifications.

- The model must include a 3D view export, with the appropriate filters applied for coordination purposes.

- Not all conflicts identified by clash detection software necessarily indicate design coordination issues. Each conflict should be assessed on a case-by-case basis and communicated as needed.

- All task teams have the responsibility to utilise point cloud survey information and conduct their own validations to ensure that their proposed designs fit with the current conditions of the building.

- The lead designer should identify design coordination issues between models from different authors, but all authors are individually responsible for the quality, suitability and spatial coordination of their own content.

- It is essential to establish the necessary rules and tolerances between various disciplines, as outlined in the BIM Execution Plan.

- At agreed timescales, generate and disseminate a report through the CDE, outlining the unresolved design issues between each discipline. This keeps all parties informed and accountable for the number of outstanding issues.

- Consider the process for coordinating highly sensitive assets in the design, taking into account the client's security requirements. In some cases, certain assets may not be represented in the model with their actual shape and data, to allow for security considerations.

- Ensure that the location of each clash is easily identifiable, making it straightforward for teams to locate and address specific clashes during the coordination process.

7.3.   No construction until completed design

There are various reasons why a design might not be ready on time, but the focus here is a recurring issue within the industry: initiating on-site construction work without a fully coordinated and completed design.

The principle of possessing a coordinated and completed design before commencing construction not only makes sense but is crucial for a project's success. However, under certain circumstances, and owing to specific project pressures, this principle can appear more like an unreachable ideal, prompting the adoption of high-risk approaches.

Beginning construction without a fully completed, authorised and accepted design brings about numerous challenges, including but not limited to, delays, budget overruns and continuous rework. Such hasty decisions can lead to confusion among team members, resulting in a detrimental loss of trust between clients and the delivery team members.

Addressing the culture in the industry

The culture in the construction industry needs to pivot from aggressive timelines that urge teams to start on-site work immediately after acquiring site possession. We need to shift towards a mentality that prioritises comprehensive surveys and detailed coordination with various task teams, minimises impacts on the project's critical path and decreases unexpected rework on site.

Regrettably, the 'shovel-ready' mindset still remains in the industry, perpetuating the perception that starting construction at the earliest opportunity will expedite the project. Some even go as far as suggesting that design details can be finalised on site, contributing to a risky 'build now, design later' approach.

Initiating construction before a design is fully coordinated can trigger a domino effect of reworks and delays, impacting the project's cost, final delivery date and overall client satisfaction. Moreover, this approach can also entail legal repercussions and potential safety hazards that might pose a risk to the workers and the public.

With people paying closer attention to return on investment and proper information management than ever before, 'shovel-ready' shortcuts are increasingly risky (Norton, 2021). These can lead to inefficiencies and delays, disrupting the 'golden thread' of information. Instead, it's far better to be fully prepared and planned, as this helps to mitigate potential risks during construction.

Therefore, it is crucial that teams follow the correct processes for information approval and refrain from using preliminary or rejected construction issue information. There have been cases in which teams have resorted to using preliminary revisions for on-site work without waiting for the required contractual construction issue to have reached the published state. Deviations from the correct process should be categorically discouraged, with practices such as 'build before design' strictly prohibited and strongly penalised, owing to the potential severe impact on businesses and the industry as a whole.

By investing extra time to complete the design and reduce potential on-site risks, we can expect substantial benefits during the works' execution, including the avoidance of on-site delays, enhanced quality and a more cost-effective project. This requires conducting necessary surveys, coordinating designs with all involved parties, and effectively communicating with clients. Remember that the ultimate goal for the client is to receive the final product promptly without compromising on quality. A comprehensive understanding of this objective can help prevent the unnecessary rush that leads to compromised quality and delays caused by rework.

To accomplish this, there is a pressing need for the industry, including the appointing party, to shift away from the mentality of commencing on site as early as possible, despite not having the completed and authorised design in hand.

While BIM may not resolve all construction challenges, it significantly assists in managing information throughout the design and construction phases. It furnishes concrete evidence of information quality and maintains records of authorised and accepted information, ready for construction, based on feedback from relevant stakeholders.

Delivery teams often find themselves under pressure from senior leadership or clients to start work on a site as soon as they acquire possession, driven by concerns around cash flow, market positioning or political factors. However, this haste can lead to significant issues if there hasn't been clear communication regarding the critical importance of completing the design phase first. The temptation to kick-start work early for perceived financial or temporal advantages should be resisted. It's vital to consider the potential risks associated with starting work before completing the design phase, and to nurture open, effective communication with the client. By focusing on quality and proper planning, rather than just speed, the industry can ensure that projects are successfully, safely and punctually completed, thereby maintaining the integrity and trust vital for a thriving construction industry.

7.4.   Are digital construction roles needed in your business?

The introduction of the BIM Mandate in 2011 (Cabinet Office, 2011) popularised new roles in the fields of architecture, engineering and construction. These roles, such as BIM manager, BIM coordinator and BIM architect, were created to assist businesses in adopting and using BIM in their day-to-day operations.

However, nowadays things are a bit more confusing. We're seeing a shift from using the term 'BIM' to 'digital' or 'digital construction'. This change has led to a whole host of job titles to describe people who do similar things, creating confusion among professionals and organisations about the specific roles, responsibilities and expectations associated with these positions.

It's important to note that some of these roles can be challenging to define, as their functions and responsibilities may vary depending on the specific characteristics of the business and the field of work. Additionally, the names used to refer to these roles can also differ depending on the company or organisation. Despite these differences, all of these roles are linked to the BIM process and the use of new technologies in the construction industry.

If we consider the role of 'digital construction lead' as an example, the responsibilities and the scope of this position will differ between a construction company and an architectural or engineering firm. Regardless, holders of the role in both organisations should possess a solid understanding of the UK BIM Framework and how this would impact their daily tasks within the structure of the business.

In my opinion, the role of a digital construction lead involves driving the company's transition from traditional methods and procedures to ones that align with the UK BIM Framework. This includes providing necessary training to upskill existing roles and provide support to clients, design and delivery teams to ensure compliance with the correct process, having appropriate documentation in place, and actively seeking opportunities to improve the design and delivery of projects using the latest technology and best practices.

To answer the question posed in the title: yes, your business will require digital construction roles, although the size of the team should be proportionate to the size of your business. It's not realistic or sustainable for a business to have a large number of digital construction specialists. Instead, I view digital construction as a means to improve existing processes and as a natural evolution of the responsibilities of existing roles within companies. I see a digital construction lead as an internal consultant who is responsible for educating teams, bringing them up to speed and guiding them to implement the correct processes, based on the latest standards and best practices in the industry. Additionally, this role should create a strategy for introducing innovation and the appropriate tools in the business.

In my view, if the digital construction lead does an effective job, the role will naturally evolve and eventually no longer be required as everyone within the company should know how to deliver projects according to the UK BIM Framework and comply with the business's digital construction strategy. However, owing to the continuous development of standards, lessons learned and new technological opportunities, a digital construction lead will always be needed within a business, albeit with a different name and different responsibilities. The role is one that is always in transition; while it may initially focus on changing the company's culture, strategy, processes and internal and external education, it tends to evolve towards utilising data, fostering innovation and researching new opportunities that technology can support within different departments of the business.

The digital construction lead must possess robust leadership skills. This includes the ability to work independently with a high level of initiative, engage with employees of all seniority levels, collaborate across different departments and interact effectively with clients. It is also necessary to have enough resilience to handle resistance from different teams. The digital construction lead should work with different departments to ensure that the innovations made fit with the company's overall strategy to enhance efficiency and productivity; these innovations should also take into account the interactions with other tools that are already in place, as well as security restrictions.

In summary, while digital construction roles are important, in my opinion, the focus should be more on providing consultancy services by supporting and upskilling other departments in evolving and delivering projects, rather than solely being responsible for day-to-day activities. The digital construction lead should play a strategic role, being responsible for driving the digital transformation of the company, aligning the UK BIM Framework with the company's strategy, and continuously looking for new opportunities to improve the design and delivery of projects with the latest technology and good practices.

7.5.   Be careful with the BIM certificates

Since the introduction of the first BIM Mandate in the UK, various certifications for individuals, organisations and, more recently, software providers have emerged, aimed at demonstrating proficiency and adherence to BIM standards. These certifications initially began with the PAS 1192 standards (BSI, 2013) and have evolved to include the ISO 19650 standards (BSI, 2019, 2020a, 2020b, 2021, 2022a).

Obtaining business certification can offer a variety of benefits to companies. One of the most significant advantages is the clear marketing benefit it offers. By showcasing their BIM certification, businesses can set themselves apart from competitors and demonstrate their commitment to following BIM standards. Moreover, obtaining business certification can boost clients' confidence in a company's ability to meet project requirements. By showcasing compliance with industry standards, companies can prove that they possess the knowledge and skills necessary to handle BIM projects efficiently.

Another key advantage of earning business certification is that it ensures that the business has the appropriate documentation and internal processes in place to comply with industry standards.

This adherence to standards can enhance the company's processes and documentation, instilling greater confidence that they are following the best industry practices. The annual audits are a crucial part of the certification process, serving as a beneficial tool to ensure that the teams adhere to required processes, thereby preventing their projects from being a source of non-compliance during auditor evaluations. These audits are instrumental in promptly identifying and rectifying any non-conformances. By indicating areas of non-conformance with industry standards, they enable companies to improve their processes and mitigate the risk of losing their certification.

Having participated in several business accreditations, I endorse these certifications because they have enabled me to review and enhance our processes and documentation in the past, instilling confidence in the validity of our work. These certifications have also provided us with a framework to follow and assisted us in staying up to date with the latest developments in the BIM industry and best practices. However, I believe that while these certifications can inspire confidence in a company's BIM capabilities, they do not guarantee that the services will meet clients' expectations, nor do they necessarily ensure that the company follows the correct documentation and processes to deliver project needs. Therefore, businesses need to be evaluated on an individual basis, irrespective of their certification; while certification can be useful as guidance, it should always be approached with caution based on the project requirements.

Relying solely on a company's accreditation is not sufficient when searching for companies that support your digital construction ethos and needs. While a certificate can provide some assurance of a company's commitment to delivering high-quality projects, it doesn't necessarily indicate that the entire company follows the same approach, or that all individuals have the same understanding or competencies. Therefore, it's crucial to conduct due diligence by interviewing the company to understand its culture and methodology to projects. This includes its values, mission and objectives, as well as its experience in digital construction projects. In this way, you can evaluate the company's compatibility with your digital construction ethos and determine whether it's the right fit for your project.

It's also important to know which team members will be working on the project and their experience in past projects, including their roles and responsibilities in the digital construction process. This ensures that the right people are assigned to the project and have the necessary skills and knowledge to complete it successfully. Throughout the project, it's essential to monitor the company's performance, track progress, identify any issues or roadblocks and address them promptly. This ensures that the project is completed on time, within budget and to the highest quality standards. Additionally, feedback gathered during the project can be considered when selecting future project partners.

Typically, a selected small team of specialists, or sometimes an external consultant, leads a company through the certification process. For successful implementation of digital construction in a project, it's crucial to establish a solid educational foundation among all relevant members of the business. This involves providing training and education on digital construction processes and tools, as well as promoting a culture of innovation and continuous improvement.

There are occasions when teams, initially proposed and interviewed during the tender stage, change after the contract is awarded or during the project's execution. The incoming team might not have the same skills or capabilities that originally led to the selection of the initial team. Therefore, it's essential to partner with companies that have shared their knowledge across all team members. While certificates can offer assurance that processes and documentation are in place, they don't guarantee successful implementation. Both the company's culture and its team members must wholeheartedly adopt the process for the implementation to succeed.

While I have acknowledged the advantages of certification, I do have reservations about the expenses involved in acquiring them. The fees charged can be too high for some business or individuals who are keen on learning more about the standards and demonstrating their skills but whose employers cannot afford the training costs. While some organisations and people are willing and able to pay a premium to add certification to their resumes or websites and showcase their abilities in an industry lacking in expertise, I believe that the high costs might exclude otherwise qualified candidates from pursuing these credentials.

Furthermore, I think that individual certification can create unrealistic expectations. In the same way that business certification can instil confidence but each enterprise must be evaluated on a case-by-case basis depending on the project needs, some individual certificates also necessitate careful consideration. While they provide a solid understanding of the process, I do not believe that they equip individuals with the expertise and skills needed to implement and lead projects effectively in real-world scenarios.

Offering training certificates to individuals and businesses serves a valuable purpose in the industry by helping them ensure that their knowledge and practices conform to industry standards and regulations. However, these certificates are superficial, and ongoing learning and development are necessary for both individuals and businesses to guarantee compliance and best practice at levels and departments. This will ensure that the certificates have real significance beyond being mere marketing tools.

I do not criticise business or professional certificates, as I recognise their value for specific needs and acknowledge there is a market for them. I also believe that further and accessible education and training are critical in supporting our industry's adoption of the digital transformation. Nonetheless, I am uncomfortable with the monetisation surrounding it, as some individuals may be misled by the certification of a business or professional, seeking recognition rather than knowledge.

The most important advice I can offer is for everyone to find the training that best suits their role and organisation and to address any gaps in their development that they consider necessary for their career. Above all, ensure that you share your knowledge with those around you and provide opportunities for them to upskill.

As previously stated, the adoption of digital construction involves a shift in business culture and the enhancement of skills among all personnel engaged throughout the project's lifespan. I am of the opinion that no certification can fully encompass these requirements.

7.6. Consider the use of external BIM consultants

The successful implementation of BIM requires all stakeholders to have the necessary expertise, knowledge and attitude, if they are to work collaboratively and efficiently. This is particularly important for companies with design responsibilities, as they play a critical role in the BIM process. It is, therefore, essential that companies with design responsibilities have the capability and capacity to deliver the BIM requirements in-house. This includes having a team of professionals with the necessary BIM expertise according to the UK BIM Framework and access to the necessary software and tools, based on their role. Failure to have these capabilities can lead to complications and inefficiencies during the project, which can increase costs and delay project completion.

As such, the selection of the task teams for a BIM project must prioritise their ability to meet the BIM requirements. This should be considered as a non-negotiable criterion, alongside other important criteria, such as cost, experience and reputation.

While cost is a significant factor in choosing a partner, prioritising cost over capability will lead to project pitfalls that outweigh any perceived cost savings at the tender stage.

Supporting and upskilling partners

From the perspective of a main contractor, there's a duty to support companies with whom there's been a long-standing positive relationship, aiding them in enhancing their skills, as discussed in Section 3.4. To this end, it's essential to offer training opportunities, such as BIM workshops, which are invaluable in helping these companies to understand your expectations as a company and support their BIM expertise. However, there comes a time when selecting the most suitable company to fulfil the project's requirements, including BIM competencies, takes precedence. If certain companies do not meet expectations, tough decisions will inevitably have to be made.

In my experience, companies that increase their fees to deliver BIM requirements can be divided into two main categories.

- The first category includes companies that advertise themselves as BIM-compliant. However, many of these companies only undertake BIM projects when specifically requested by clients. Often, they do not fully grasp the advantages and core principles of BIM. While they can produce designs in a 3D environment, they do not consistently apply the BIM process as standard practice. Consequently, they feel justified in charging extra for BIM services, viewing it more as a special service to clients rather than recognising the benefits it brings to the entire project, including for themselves. Typically, these companies are in the initial phases of their digital transition, with limited experience and a small team equipped with the necessary BIM skills. This often makes collaboration challenging, owing to their limited expertise and resources. They might also find it tough to handle unforeseen challenges or complications that arise during the project.
- The second group consists of companies without an internal design team or the essential skills to produce a 3D design and relevant asset data based on project specifications. In such situations, their only recourse is to engage a BIM consultant. For them, the BIM process is merely an added expense, which they pass on to the client, without considering the potential benefits for their own business.

It's worth noting that neither category necessarily offers value for money. Companies with a genuine digital transformation ethos, treating BIM as routine, often complete projects more efficiently and cost-effectively than those that require external support.

The role and value of BIM consultants

BIM consultants can be an effective option for companies embarking on their digital transformation journey or those needing guidance to meet project requirements. BIM consultants can address skill gaps in the early stages of a company's transformation, offering invaluable advice to propel the business's overall digital evolution. However, it's crucial to recognise that routinely relying on a BIM consultant to fulfil project requirements might diminish competitiveness, raise costs and extend delivery times. Unless they're thoroughly integrated in your team, it's wiser to employ them to steer the transformation of your business or as supplementary support to your in-house BIM processes, rather than as a stand-alone delivery team that recreates information in the necessary format.

Selecting a BIM consultant

If you are seeking a BIM consultant to deliver the project requirements, akin to the point cloud survey consultancy mentioned in Section 7.8, it is essential to establish partnerships with the right companies. Choosing the right BIM consultant, one who has an educational and collaborative approach with your business, can make a significant difference in the success of your project and digital transformation journey. It is important to note that some consultants charge exorbitant fees for certain tasks. Hence, it is crucial to understand what you are purchasing and know what a fair price is when engaging with BIM consultants. Regrettably, in the industry, these excessive fees have contributed to the negative reputation that BIM has acquired.

Therefore, before choosing a BIM consultant to assist in delivering the project BIM requirements and support you with the digital transformation within your business, if you don't have in-house resources, it is crucial to conduct thorough due diligence to understand different consultants' capabilities, capacities and approach.

From a client's perspective, when requesting documentation to support the tender process, I have come across consultants who seem to prioritise quantity over quality. These consultants tend to produce unnecessarily large and complex documentation for BIM implementation instead of delivering high-quality, relevant content for the project and client needs.

Such an approach can make an already complex process even more challenging for the industry to adopt and might not yield the expected results. Thus, it is essential to choose a BIM consultant who values quality over quantity and simplifies the process for the organisations involved.

Evaluating cost and BIM competency

Following up on previous discussion, the choice of a task team for a project often hinges on cost. Many companies are tempted by the cheaper option. However, if the chosen company cannot deliver BIM, there might be additional fees to meet the project's requirements with a BIM consultant's assistance. This can make the cheaper company, the one without BIM capabilities, more expensive in the long run, and potentially complicate the delivery process. This contrasts with other companies that readily offer BIM. Thus, when deciding which company to select for a project, it is vital to weigh these factors carefully. This scenario underscores the importance of upskilling the different roles within task teams, akin to the discussions in Section 2.7 regarding upskilling within main contracting firms.

By enhancing the BIM competency of these teams, companies can potentially avert the hidden costs and project delays associated with outsourcing BIM expertise, ensuring smoother and more cost-effective project execution. Additionally, where companies might not have the financial means to support internal BIM consultancy resources in permanent roles, engaging external BIM consultants can be a viable strategy to support this upskilling initiative. These consultants can provide the necessary training and guidance to elevate the BIM capabilities of the task teams, fostering a more self-sufficient and competent workforce that can proficiently navigate BIM-related tasks in future projects.

There may be situations where a task team without BIM capabilities is appointed, possibly because of novation. In such cases, it's crucial to engage a BIM consultant to ensure that the project's requirements are achieved. Before moving forward, it's essential to define the relationship and agreement between the task team and the BIM consultant to avoid duplication, which can negatively affect the design programme. The chosen approach should suit the project's scale and intricacy, keeping time and budget in mind.

The conventional method involves the design consultant or the subcontractor providing a set of 2D design information, which the BIM consultant then converts to a 3D model. This approach doubles the design task, heightens the possibility of errors, raises the costs and could considerably delay the design programme. It's also crucial that all drawings, schedules, schematics and other documents delivered to the common data environment stem from the model and not the 2D drawings created by the design consultant or subcontractor.

Challenge mitigation

To reduce duplication and accelerate the design process, it might be more economical and efficient for the BIM consultant to design the project directly in a 3D environment with input from the task team. This way, there's no need for duplication, and the design process can move forward without hitches. No matter the design approach chosen, it's essential to allocate a sufficient budget for the BIM consultant to modify the model based on agreed alterations during the design and construction phases, ensuring that it reflects the actual work done on site.

As highlighted, when choosing companies that don't have internal BIM expertise, it's crucial to ponder the possible repercussions on the design programme, budget and overall project timeline. Therefore, partnering with a BIM consultant to formulate a robust strategy that meets the project's needs is essential. Adopting this strategy can help companies reduce risks, ensuring that they deliver top-notch construction projects on time and within budget.

Although the initial costs of enhancing skills and integrating new processes and technologies might appear steep, the long-term advantages are considerable. Such investments can result in more precise and efficient project execution, enhanced communication between stakeholders and, ultimately, heightened business success. As BIM practices gain traction, companies that don't invest might find themselves lagging in the competitive race for future projects. In the grand scheme of things, investing in skill development and embracing BIM practices can enhance competitiveness, enrich project results and bolster the industry's reputation, setting companies on a path to long-term success.

7.7.   Challenges when dealing with existing buildings

Implementing the BIM process for existing buildings presents distinct challenges that differ from those encountered in new construction projects. When working with existing buildings, it is important to consider the following factors.

Dimensional survey

One of the main challenges is the lack of accurate legacy data about the existing building. There is a clear need to conduct the necessary surveys and gather the necessary information to develop an accurate design for the project. This is especially true if the building is old or has undergone repeated renovations or additions, as the available documentation might be incomplete or outdated. Additionally, there may be limited access to the building for surveying purposes, either because the client does not have the lease and cannot allow intrusive works on site or because the building is still occupied and surveys might not be allowed. This can further hinder the ability to gather accurate information about the building and create a comprehensive design at preconstruction stage.

In relation to this topic, there is another obstacle that might arise when implementing BIM on existing buildings. This obstacle involves matching new services with existing ones that may be retained during the construction phase. This situation can be particularly challenging in projects where services are concealed and a complete strip-out cannot be performed before finalising the design phase. This scenario can lead to an increased number of assumptions in the design, resulting in uncertainties and potential complications during construction. To avoid this, it is crucial for the lead appointed party to conduct comprehensive surveys as promptly as possible. The goal is to gather accurate information about the existing building and its infrastructure in order to eliminate any assumptions in the design. This ensures proper coordination in the design between the new components and the existing ones, minimising the likelihood of complications and ensuring smooth progression of the project.

Section 7.8 is dedicated to 'point cloud surveys' and extensively covers the challenges involved in validating design information and existing building conditions. Through a detailed exploration of this surveying method, I delve into the intricacies of the process and provide comprehensive guidance on the key considerations when conducting a survey of an existing building.

Validation and verification survey for mechanical and electrical (M&E) services

It is crucial to communicate clearly with the client regarding the limitations of capturing data for existing and retained services. The information provided to the facility management team might not be as comprehensive for existing services as it is for new elements. A point cloud survey will accurately capture location and dimensional information for existing services, but additional details, such as condition, compliance, capacity, performance and integration with the new design, require the support of an appropriate M&E validation survey. This is necessary to identify any problems and incorporate them into the specification early on. It is important to note that, even with the right validation and verification survey, the data provided for existing services will be limited compared with those available for new services.

However, when the facility management team has a well-documented existing asset register, this can serve as a valuable resource for generating a more complete dataset, encompassing both existing and new elements. This integration of data bridges the gap between the old and the new, facilitating a comprehensive understanding of the entire system for the facility management team.

To effectively manage client expectations, transparent communication is essential. Clearly, conveying any limitations or constraints in the data helps avoid misunderstandings or disappointments. Clients should be fully informed about the challenges associated with capturing data for existing services.

Fast-track programme

When dealing with existing buildings, it is often common to come across fast-track programmes that require swift design and construction stages. In some cases, this can result in the overlapping of design and construction phases. The accelerated pace poses challenges in terms of thorough project design review, stakeholder coordination and effective information management. Often, teams might bypass the usual approval process to keep pace, but it's essential to remember that proper team engagement in following established processes is crucial. Shortcuts in managing information are unacceptable because of their costly long-term consequences.

Bypassing the correct design approval process can result in significant issues later. These are often more expensive to rectify than any initial gains from rushing. Hence, strict adherence to the information management process is vital in controlling costs and maintaining quality.

To mitigate complications, it's crucial to establish clear communication channels and a strategy for managing design information without compromising project quality. Encouraging strict adherence to procedures and high team engagement can minimise potential issues and ensure that the project stays on track. This approach promotes understanding, reduces the risk of costly errors and drives project success.

Mindset barriers

There is a common misconception that BIM is only useful for new constructions, causing some to resist its application to refurbishments, extensions or fit-out projects. However, with the right approach, BIM can be beneficial for existing buildings too, providing numerous advantages to the project team. Detailed information on the current state of a building allows the team to anticipate issues, improve coordination and decrease construction errors. Moreover, BIM can enhance communication and ensure that everyone has access to identical information, minimising misunderstandings. Implementing BIM on existing buildings may present additional challenges, compared with doing so on new construction. However, as we've observed, these challenges aren't necessarily specific to BIM implementation; we would encounter the same challenges in non-BIM projects. They arise primarily from the difficulty of obtaining accurate information about the existing building and its retained services or structural elements.

Additionally, they're often a result of the industry's cultural tendency to start on-site work as quickly as possible, rather than properly planning the work and avoiding shortcuts to deliver the projects, as thoroughly discussed in section 7.3. It's, therefore, necessary to adhere to proper procedures for all projects, whether they are BIM or not. This involves conducting necessary surveys, developing a well-coordinated design that considers all pertinent factors and minimising assumptions. Furthermore, it's crucial to manage information properly, in accordance with the mutually agreed data protocols in the common data environment.

In my view, it is preferable to refrain from hastily initiating on-site work, only to later confront unforeseen issues. Instead, comprehensive planning and verification of tasks should be carried out before starting on site. While this method might demand more initial time investment, it is ultimately more cost-and time-efficient for all involved parties. Additionally, this leads to an improved client experience in the long term.

The process

Let's consider a typical scenario that often occurs when working on existing buildings. When the project kicks off, assuming we already have some preliminary information about the existing conditions of the building, we start developing the design in a 3D environment, using the available information about the existing building conditions. Normally, this information is in two dimensions. However, as a standard practice, this preliminary information provided by the client must be checked and cross-verified using a point cloud survey to accurately map out and measure the physical space. This process confirms the validity and accuracy of the available information, ensuring that the design is being developed correctly. But we need to bear in mind a crucial point: during the point cloud survey, only exposed elements will be captured; therefore, we can only validate and verify these visible elements. There may be certain elements, perhaps concealed or obscured, that won't become apparent until the site has been thoroughly stripped out. These hidden factors, once revealed, could impact the overall development and coordination of our design.

Therefore, when the client assumes control of the site and gives the go-ahead for more intrusive work, the site team steps in to undertake the strip-out of the site, methodically exposing any elements that were previously hidden. It is crucial to consider these elements as part of the design process, as they will inevitably have some degree of influence on the construction stage. On completion of the strip-out, revealing all relevant elements, the survey consultant performs a second point cloud survey. This repeat scan gives us an even more comprehensive understanding of the site and its nuances, helping us make better-informed design decisions.

The survey consultant and project delivery team then come together to assess the results of this second survey. They collaboratively interpret any additional information gathered, carefully evaluating its potential impact on the project.

Once any discrepancies or conflicts between the developed design and discoveries made during the strip-out have been identified, the design is updated to incorporate the new findings and resolve any potential new design conflicts before completing the construction issue information.

The overarching objective of this complex multistep process is to verify and enhance the design as much as possible. We do this by conducting thorough and relevant surveys and by minimising the amount of guesswork or assumptions in the design process. This approach allows us to identify and resolve any design issues in advance, ensuring that the entire team is fully cognisant of the task ahead. This, in turn, prevents potential delays during the construction phase.

Figure 7.5 clarifies the process of implementing BIM on existing buildings. It aims to demystify common concerns and obstacles associated with integrating BIM in already-constructed buildings.

Figure 7.5 Implementing BIM on existing buildings

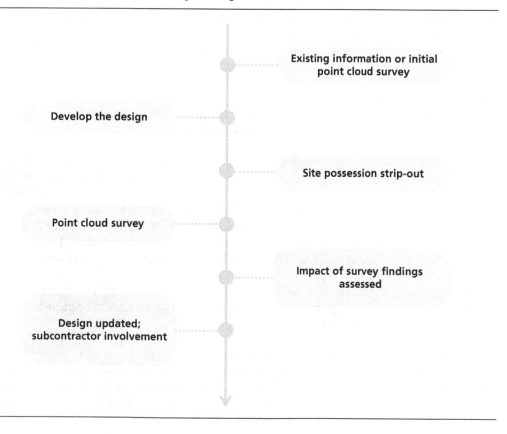

205

7.8. Point cloud surveys

In my experience, a point cloud survey can prove invaluable for any company involved in new or existing building projects. This survey holds particular significance for refurbishment and extension endeavours, as it offers detailed information regarding the current condition of a building. This information greatly aids the design team in developing the appropriate design during the preconstruction stage. Moreover, the survey acts to validate and verify the design information and allows assumptions to be discarded; it is thus possible to ensure that the on-site work corresponds to the planned construction stage and to identify any errors before they can escalate into more significant problems.

Point cloud surveys have increased in popularity in recent years, thanks to continuous technological advances, improved affordability and faster processes than in the past. While operating the scanner might not appear particularly challenging, the true value of these surveys lies in the utilisation of raw data. Not everyone possesses the skills or expertise to generate valuable outputs from a point cloud survey, making it crucial to have a specialised team in place. This team, whether they are sourced internally or externally, can conduct the survey and provide valuable output.

If your business lacks the necessary capabilities and you rely on point cloud services, it might be worth considering establishing a partnership with a reliable and trusted provider. This partnership can offer a regular service and discounted fees. By reviewing the provider's performance annually, you can ensure that you are still working with the best partner for your needs.

The reason for this recommendation is that while many companies offer point cloud services, it can be challenging, at least in my experience, to find reliable partners who truly collaborate with your business and consistently deliver information within the agreed time frames and expected quality. Considering that delivery of the results of the point cloud survey is a key milestone in project progress in many cases, it is essential to avoid delays that can affect the rest of the programme.

To effectively conduct a point cloud survey, it is crucial to consider the necessary time and cost. By planning, reviewing, validating and verifying designs in advance, as well as the completed onsite conditions, we can minimise the risk of unexpected issues during the construction stage. This ensures that projects are completed efficiently and effectively.

Initially, when approaching the market and consulting different point cloud providers, you might encounter challenges in obtaining accurate and comprehensive quotations for point cloud surveys. The fees charged by different companies can vary significantly for the same project. Hence, it is crucial to be specific and clear about the information you require when seeking a quote. Having a clear understanding of your goals and desired outcomes from the survey will assist in determining the most suitable approach and ensuring that you collect the necessary data. This section also addresses the considerations you need to be aware of when defining the scope of work.

It is important to note that the cost and time implications for conducting the appropriate survey during the preconstruction stage are minimal, compared with the benefits to be gained in reducing risks during the construction phase of the project.

Another key aspect to ensure the successful implementation of the point cloud survey is the cultural change required within your business. It is common for the commercial team to initially perceive the survey as a cost, but it should be viewed as an opportunity to enhance the design and construction stages. Therefore, it is essential for teams to understand the opportunities and benefits that conducting a point cloud survey can bring to a project.

Figure 7.6 Point cloud survey approach

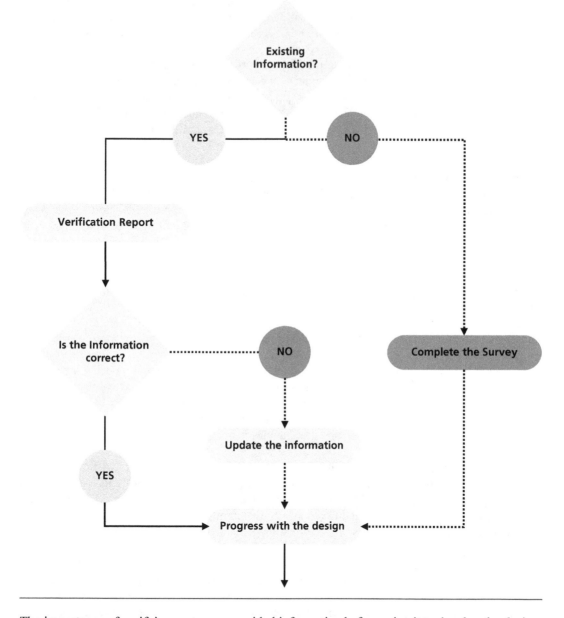

The importance of verifying customer-provided information before using it to develop the design, as well as the impact of conducting a point cloud survey when no existing customer information is available, is illustrated in Figure 7.6.

Why you need a point cloud survey

Point cloud surveys are immensely useful and have extensive applications across a variety of projects, not limited to existing buildings, as some people might assume. Their utility is extended to civil projects and new constructions as well. For infrastructure assets, such as bridges, tunnels or roads, these surveys provide a high-definition snapshot, capturing precise data; this is crucial for structural analysis, repair, maintenance planning and refurbishment projects. In new building projects, they assist in accurately understanding the site topography and surrounding structures, ensuring that the new building's design complies with regulations and harmonises with its intended location. Therefore, the principles covered in this section would be applicable to these kinds of projects too.

During the design phase, it's common to encounter projects hampered by poor-quality information provided by the client at the tender stage. This information may encompass paper drawings, 2D CAD, 3D models and specifications. Often, the inadequacy of this information stems from either its initial poor quality or a lack of maintenance over time. Therefore, it's important to address this risk by verifying the information's quality before using it to develop the design.

A similar scenario occurs when receiving information from the client's current or previous design team. In such cases, the accuracy of the design information cannot be trusted, owing to a lack of control over its development. Hence, it is necessary to conduct an appropriate survey to validate and verify any information provided by the client. When tasked with design responsibilities, it is not advisable to assume the accuracy of designs completed by others.

In summary, during the design phase, you face two scenarios: the need to conduct a point cloud survey to start the design from scratch, owing to unreliable existing information, or the need to conduct a point cloud survey to validate and verify a design completed by others.

There is a common misconception that a point cloud survey is necessary to start a BIM project. However, this is not entirely true. While I do recommend completing a point cloud survey at the beginning of the design process, owing to the likelihood of existing information being inaccurate, there are situations where project constraints, such as restricted time frames, client restrictions or limited accessibility, mean that the design is begun based on existing information. In such cases, it is important to pay special attention to the risk of inaccuracy and be prepared for the possibility of additional work to address any significant inaccuracies in the existing information.

As soon as possible, the point cloud survey must be completed to verify the existing condition of the building against the design and make necessary updates. This approach applies to both 2D CAD designs and BIM projects. Therefore, it is essential to understand each project's limitations, explore different types of survey and find ways to overcome barriers in the design development while considering the need to verify any information generated.

During the construction stage, the approach is different. You may need to verify the work completed on site for specific trades, to ensure accuracy and adherence to the authorised and accepted design. This helps avoid issues later on that could cause problems with other trades and the overall project execution. Therefore, it is essential to appreciate the project's requirements and limitations, understand and explore different types of survey and find ways to overcome barriers in the design development, while being mindful of the need to verify any information provided, as well as to verify the works completed during the construction stage.

It is important to recognise the significance of conducting the necessary survey at the earliest possible stage, as this greatly contributes to the successful execution of the overall project. Therefore, it is essential for the client to understand the importance and benefits of completing the survey, as well as the associated time and cost implications during the preconstruction phase. By addressing these considerations, the risks during the construction stage can be minimised, resulting in a better project execution that benefits all parties involved. It is also important for the client to actively support and facilitate the completion of the survey by providing access to the building. Moreover, it is worth noting that the cost and time implications of conducting the survey are relatively minimal, compared with the potential issues that can arise during the construction stage if the necessary surveys are not conducted in a timely manner.

Point cloud surveys compared with traditional surveys

In essence, a point cloud survey enables the capture of a larger volume of data in a shorter time frame, leading to more comprehensive and accurate information. This reduces the necessity for site revisits and improves overall efficiency. Point cloud surveys utilise 3D data points, resulting in highly precise and detailed representations of existing conditions for buildings or sites.

Additionally, modern scanners employed in point cloud surveys can produce high-quality images, which can be valuable for dilapidation surveys, eliminating the necessity for a separate task.

It's important to note that the benefits of point cloud surveys are applicable to both BIM and non-BIM projects, although the specific type of survey and outcomes will depend on the project's requirements.

Process for existing buildings

I want to expand on the process that we covered in Section 7.7, on how to use a point cloud survey on existing buildings, as this is the scenario where there are more options and opportunities, and where teams tend to get confused.

In a case where you have not received any design or legacy information from the client regarding the existing building conditions, a point cloud survey becomes the only viable option to facilitate a viable design process. This survey accurately captures the dimensions of the building, minimising assumptions and associated risks in the design process.

It's crucial to note that if you perform a point cloud survey before the strip-out phase to obtain key building dimensions, you will need to repeat the survey after the strip-out. This ensures that any discoveries of previously hidden items are taken into account. Point cloud surveys can only capture visible elements, so features concealed by ceilings or plasterboard will remain unrecorded until they are exposed. Incorporating these discoveries in the design is essential.

When a client provides legacy information about a building as part of the existing and historical data, it is important to validate and verify this information. This becomes particularly critical when you assume design-and-build responsibilities for the project. Since you will be held accountable for the design, ensuring that the provided information aligns with what needs to be developed and built is essential. Validating and verifying information from both consultants and clients helps identify potential risks that might arise during the construction stage.

As mentioned earlier, legacy information is often inaccurate and fails to reflect the actual conditions of the building, which are needed when developing the correct design. I have encountered situations where the provided drawings had incorrect scales, misplaced partitions, incorrect thicknesses and similar issues with services.

During the verification survey, which involves comparing the point cloud data with the information provided by the client, a report is generated to identify any discrepancies and assess their impact on the project. This design verification becomes especially important when the client's requirements include a specific net internal area (NIA) and minimum ceiling height restrictions.

If discrepancies are identified and updates are necessary, it is important to consider the impact on the project timeline and communicate this to the client. If the existing information or the design developed by the client team is correct, the project can proceed as planned.

Types of survey and their purposes

While there are various types of point cloud survey and outcomes with different purposes, I will discuss the ones that I have encountered in previous projects as they tend to be the most common in the construction industry.

Handheld survey

A handheld or mobile survey is used to gather basic information about the existing conditions of a building when access is limited. This approach is generally fast, less disruptive and less expensive.

Handheld surveys are best suited for situations where only basic information is required and access is limited. If the survey requires more detailed or accurate data, this option might not be the best fit for the project. While the information provided is useful, it might not be as accurate as a terrestrial point cloud survey. Handheld surveys are often employed when a client hasn't provided any existing information about a building and basic knowledge is needed to develop a design while the building is occupied.

Terrestrial point cloud survey

Terrestrial or static point cloud surveys utilise 3D laser scanning to digitally capture the shape, dimensions and other physical characteristics of structures and objects. This is achieved by directing a line-of-sight laser beam at the object being scanned. While new scanners allow for automated mapping, this is still a slower process compared with a handheld survey. However, the accuracy is higher, making it perfect for jobs where each millimetre is important.

One challenge with point cloud surveys, including handheld surveys, is that they can only capture visible objects. Therefore, it is crucial to communicate clearly any requirements to capture objects or features that might be hidden from view, such as services above ceilings or steelwork. By considering these factors, you can ensure that your point cloud survey is as accurate and comprehensive as possible.

Drone survey

Also known as UAVs (unmanned aerial vehicles), drones are used to conduct surveys and collect point cloud data from an elevated position. Although they might not be as accurate as terrestrial point cloud surveys, drones are highly effective in scanning large areas, making them valuable for surveys involving façades and land.

Scan-to-BIM

Scan-to-BIM is the process of converting a point cloud survey into a 3D model. The goal is to create a highly accurate and detailed digital model that captures the existing conditions of a building and can be used for design development.

It's important to note that the time and cost of the scan-to-BIM process depend on the project's complexity and the amount of data to be processed. This should be considered at the beginning

Figure 7.7 Site verification using point cloud data

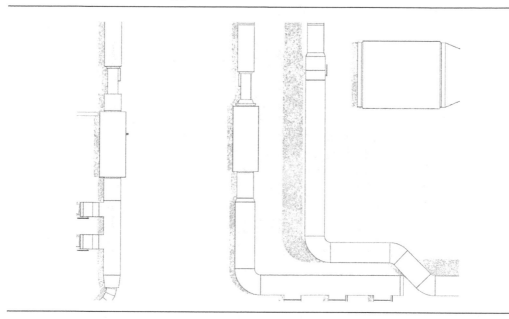

of the project. The point cloud information can additionally be converted into 2D information, if required by the project.

Site verification

Site verification, through a point cloud survey, involves comparing the point cloud data with 2D and 3D design information. As shown in Figure 7.7, the 2D or 3D design information can be overlaid on the point cloud survey information, providing a visual representation that aids in identifying and understanding discrepancies between the actual site conditions and the location of the ductwork given in the accepted design.

This report facilitates the implantation of corrective measures, either before the onset of construction or before other subcontractors are affected. Any discrepancies or errors could significantly impede the construction process, potentially leading to delays or additional costs. Therefore, verifying their correctness is of paramount importance. Utilising a point cloud survey for the verification report can enable the early identification of potential issues. This, in turn, allows for proactive measures to be taken to mitigate any adverse effects on the overall progress of the construction project.

I would like to clarify that the verification report will only cover elements that are visible when the point cloud survey is completed. If you are requested to produce a verification report at the end of the project for handover, as some clients have required in the past, please make them aware of the limitations of conducting such a report at that stage. Conversely, compiling a verification report throughout the construction process can be quite expensive. Therefore, I recommend conducting the point cloud verification survey at specific milestones, focusing particularly on work packages requiring precise execution, to prevent disruption to other trades. The millimetric accuracy of point cloud data is crucial in these instances. For regular weekly verification, where accuracy is less critical, the alternative approaches discussed in Section 4.6 may suffice, as this level of precision is not necessary. These methods will provide adequate data to confirm that the work has been carried out correctly.

Challenges

Here are several common hurdles that you might encounter when conducting a point cloud survey.

Site accessibility

Early arrangements for the surveyor's access to the construction site are crucial to prevent unnecessary delays in the project timeline. This involves ensuring entry to all requisite areas, such as rooms, elevator shafts and risers, and may also require the removal of ceiling tiles or other fixtures for the survey. Having personnel on site to aid the surveyor and ensure full access and adherence to all requirements is highly beneficial. Failure to manage these steps properly could result in an incomplete survey, missed important details or the necessity for a follow-up visit, wasting valuable time.

Strip-out process

A point cloud survey before a building's strip-out provides limited data, mostly confirming the building's boundaries. However, a post-strip-out survey offers detailed insights, revealing underlying structures that inform design decisions. Given these different benefits, two surveys, one before and one after the strip-out, can be beneficial. This approach not only gives a thorough understanding of the building's structure but also reduces unexpected issues during design or construction, potentially saving time and money.

Scan-to-BIM

While the design team may have the ability to transform point cloud data into 2D or 3D models, some people might shy away from this task, owing to the large file sizes or a lack of knowledge. This job usually falls to the surveyor. Given the time-intensive nature of this task and potential resource constraints, it can lead to delays. As such, establishing strict deadlines and monitoring progress closely is advised to mitigate potential holdups. If delays happen, it's important to hold the surveyor accountable to safeguard your design timeline. Keep in mind that some firms outsource this task overseas, which might affect the model's quality and delivery time. Therefore, it's crucial to be familiar with your supply chain before committing to any appointment.

Existing services or structures

While the design consultant usually models all new elements, the design team might be hesitant to model existing ones. It's crucial to define who will model retained services and structural items from the project's outset. In projects requiring a verification report of the design, it's also crucial to identify who will model any existing elements that are not included in the design. It's recommended to assign modelling tasks to the design consultant or subcontractor and compensate them accordingly.

I've seen situations where minor errors in the models produced by the surveyor were exploited by task teams to extend deadlines and increase fees.

Previous surveys

When a previous point cloud survey has been carried out, gaining access to the control network from that survey is vital for proper alignment of the verification laser scan with the existing drawings. However, this information isn't always readily available. Therefore, reaching out to the client and the team involved in the previous survey for this information is crucial. Lack of a control network can cause misalignment and inconsistencies between the new and old data, escalating the survey's duration and cost, diminishing data accuracy and impeding effective collaboration, owing to an absent common reference system. Therefore, it's advisable to create a control network, a set of reference points used for spatial data alignment, with Royal Institution of Chartered Surveyors (RICS) Band A accuracy that can be used by other entities needing further surveys, thus eliminating the need for numerous control networks and consultants.

Prestart and completion

Before initiating the point cloud survey, it is essential for the design and operations teams to establish a comprehensive work scope, to ensure that the data collected and the final results are beneficial to all involved parties. On completion of the point cloud survey, the surveyors should share their findings with the delivery team, address any questions and discuss subsequent steps. It's imperative not to make assumptions regarding the accuracy of the information; hence, the deliverables must be thoroughly reviewed and verified by both the design and operations teams. Engaging all parties from the outset, agreeing on a precise scope and verifying the quality of the information produced are crucial steps, but are not always taken.

Deliverables

When defining the scope of work, it is important to reach an agreement with the relevant task teams regarding the project deliverables. The following list outlines the typical deliverables that I have requested in my previous projects.

Point cloud data

The point cloud survey is a robust tool in building analysis, helping to confirm the dimensions and distinctive features of a structure. It is often used to identify any structural or design anomalies that require attention. Additionally, the data can be transformed into 2D drawings or 3D models, offering a multidimensional view of the building. This is a valuable asset in validating and verifying existing designs and site conditions and serves as a significant contributor to the construction process. For the sake of accuracy and consistency throughout the project, it's crucial that the point cloud data are integrated in the common data environment. This ensures that all firms engaged in the design process can access and utilise the data effectively.

Photographic report

Navigable photographic reports cater to users who might not have the specific software or skills to open and interpret raw data. These reports provide an immersive and interactive experience, enabling viewers to virtually walk through the building, enhancing their comprehension of the building's actual condition. During a point cloud survey, navigable photographic reports are invaluable tools for consultants and members of the design team. They provide a virtual familiarity with the building and help to identify areas of interest or concern, fostering better collaboration, as findings can be easily shared with other stakeholders.

Heat map

Heat maps, as the name suggests, are graphical representations of data characterised by a colour gradient indicating the intensity and location of the data. When it comes to understanding deviations in a structural component, such as a column, slab or soffit, a heat map visually articulates variations in plumbness, thickness, flatness or other surface characteristics. Heat maps are incredibly beneficial for swiftly pinpointing areas with significant deviations or regions that might present challenges when installing required services. It's important to remember that the colours used in the heat map can signify different levels of deviation from a specified standard or target value.

2D drawings

In some projects, 2D drawings may be required as part of the project deliverables. These should include plans at every level of the building, elevations, sections and annotations, as required. To ensure accessibility and compatibility, these should be supplied in PDF and any other file formats required by the delivery team.

Verification report

Depending on the project and information stage, it may be necessary to verify the information produced at a certain stage or verify the work completed on site.

In a verification report, a digital 2D or 3D model is compared with a point cloud survey to determine the accuracy of the digital model, as shown in Figure 7.7. A colour-coded report identifies discrepancies between the model and the point cloud survey. Green is used for items within tolerance, yellow for items outside tolerance and red for items present in the model but absent in the survey. The delivery team typically set the tolerances for comparison before the verification process begins. As a guideline, the report should highlight any discrepancies above 15 mm. The report's format allows for easy viewing in a model viewer, enabling the user to quickly identify individual items and discrepancies.

3D model

A 3D model produced in the scan-to-BIM process is a detailed digital representation of an existing building, created by processing the raw data in the authoring tool agreed on as part of the project scope. This allows for the production of a digital depiction of the captured data. The tolerance for the modelling process, as well as the level of graphical and non-graphical information for each specific element, should be agreed on as part of the project scope. The model should be developed in sufficient detail to generate 1:50 or 1:100 scale survey plans, sections or elevations, based on the project's specific requirements. If certain elements cannot be modelled within the agreed tolerance, they should be marked in the model. This helps the team in identifying objects that have been modelled outside the accepted tolerance range. The naming of objects should comply with the project's requirements and the model should be georeferenced.

Considerations when requesting a quotation

The following guidelines are intended to help you get started when discussing point cloud survey requirements with consultants. Keep in mind that these are general recommendations, and you will need to review and tailor them according to the specific needs of your project and the consultants' expertise. However, they can provide a useful starting point for your conversations with point cloud survey consultants.

To ensure that the correct appointments are made for point cloud surveys, it's important to consider various factors during the tender and planning stages of a project. Here are a few key considerations when ordering a point cloud survey.

- The time and cost of conducting the relevant survey should be determined, as well as the necessity for surveys both before and after the strip-out. The post-strip-out survey is particularly crucial for capturing all elements and minimising design assumptions when working on existing buildings.

- Ensure that the survey is treated as a critical path in the project programme and that access to all necessary areas is provided. If drones are needed, there might be specific permission requirements, based on the project location.

- Engaging design consultants and subcontractors to define the scope of work, and clarifying responsibilities for modelling and capturing retained services and structures when these are exposed, will help ensure a smooth survey process.

- When reviewing quotations from external companies, verify the information provided and ensure that it meets the minimum requirements outlined in the point cloud survey requirements.

- Arrange a meeting with consultants, subcontractors and the survey company to review information in detail. This will help identify potential issues, align expectations and ensure effective survey execution.

- For modelling production, specify the graphical information needed for each element rather than just stating a level of detail (LOD), as different elements require varying levels of graphical information.

- Finally, consider the impact of slab deviations on the design, particularly where there are ceiling height restrictions.

Point cloud

Point cloud survey accuracy will meet the following requirements:

- A colour survey for preference, but this takes longer than a survey in black and white.
- Permanent control network, set up with an accuracy of RICS Band A.
- Sufficient retro targets, on each floor, to facilitate revisits if required.
- Point cloud accuracy of Band C, as detailed in the third edition of *Measured Surveys of Land, Buildings and Utilities* (Groom *et al.*, 2014) or better.
- Point cloud density of 10 mm.

The scope of the digital survey will identify what information must be captured during the point cloud survey and will comprise:

- Elements of architectural details and structural features.
- Roof entities, such as chimney stacks or large plant.
- Sanitary features (WCs, basis, sinks etc.).
- Mechanical, electrical and plumbing (MEP) equipment (boilers, air handling units, flues, risers, etc.).
- All ductwork or pipework.
- Slab thickness, floor finishes, raised access flooring and so on.
- Railings, balustrading and forms of edge protection in default patterns.
- Additional detail to doors and windows (mullions, opening lights, etc.).
- Suitability for recording or verification surveys.
- Suitability for developing design.

Data transfer and storage

The digital survey photographic report should be securely stored by the survey company for a minimum period of 12 months before it is destroyed. A copy of the digital survey should be provided and uploaded to the project's common data environment.

Liability

The survey company will be responsible for:

- Accurate surveying using calibrated and tested equipment, with calibration data or certificates to be provided on request.
- Accurate conversion of data provided from a raw data format into the agreed file formats.

Data formats

The survey company is to process and register the cloud point data to form a complete dataset.

Coordinates

The data should be set in world coordinates (longitude and latitude); if using point cloud information, a local project base point will also need to be determined.

Verification survey

A verification survey is essential to ensure that the design provided by the client team coordinates with the existing conditions of the building and also to verify the accuracy of the completed works during construction. The aim in this proactive approach is to protect your business contractually from any inaccuracies in the design or any works executed incorrectly on site. Consider the following when requesting a verification survey:

- A colour-coded report, identifying items out of tolerance.
- A report that identifies the locations of individual items in a model viewer.
- Registered data to provide clash detection against 3D models.
- A registration error report, provided by the survey company, as part of the quality assurance (QA) process.

Model requirements

The following requirements should address the needs of the delivery team with regard to the model expectations from the survey consultant. These requirements are often used when the consultancy has not developed a design for the building and a model of the existing conditions is needed to develop the design. They may also be used when discoveries made during the construction stage are to be incorporated in the model by the survey company.

- Produce a model using the agreed software version identified in the BIM Execution Plan.
- The model shall conform to the agreed level of graphical information, depending on the purpose of each element.
- All files, including the model, shall comply with the BIM Execution Plan requirements, such as the name, metadata and size.
- The model is to be constructed to a tolerance of 15 mm of the point cloud.
- Deviations over 15 mm should be highlighted in red and communicated to the design team.
- All objects in the model should be classified as existing elements and named according to BS 8541-1 (BSI, 2012).
- The model is to be georeferenced and the project is to be set up to Project North on final delivery.
- The model must reflect any visible element captured during the digital survey.

Drawings

Drawing plans in PDF and CAD format should be provided for every level, including elevations, sections and annotations as required.

Deliverables

Based on your project needs, the following could be some of your project deliverables:

- Unified .rcs point cloud files, split into floors and downsampled to less than 2 GB.
- The full dataset including individual scans, available in e57 format if required.
- Documentation on each floor plan of the control network locations, with coordinates for each station and retro target. Ensure that these are not covered during construction if they need to be reused.
- Survey reports, demonstrating that the accuracy has been achieved and stating the error.
- Any documents named according to the project's naming protocol.
- The digital survey photographic report, with a navigable format, to be provided as soon as the scanning is completed.
- A heat map to understand deviations in the slab and ceiling, for projects where the ceiling void to install services is limited and the ceiling height is a restriction.

Other considerations

- The survey consultant is expected to visit the site and provide the deliverables within the timescale that was agreed on during the tender stage.
- Deliverables should be presented to the delivery team and any raised questions resolved.
- The operations team will provide access to all areas in the building to be scanned. The survey consultant should report any obstructions as soon as they are identified. The survey will not be considered complete until all areas are scanned.

7.9. Model requirements

The exchange of 3D models between diverse disciplines is an integral aspect of the BIM process, as it fosters a collaborative approach that is vital to the success of any project. This success is greatly influenced by the quality and accuracy of the models shared between the various stakeholders. To ensure that a 3D model is not only accurate and efficient but also easily accessible to all parties involved, it is necessary to take various factors into consideration when preparing the model for exchange.

Following the recommendations and guidelines outlined in this section will enable you to create and maintain 3D models that are correctly structured and managed. This, in turn, allows for seamless collaboration and coordination among the different disciplines involved in the project. Consequently, the integrity of the model is preserved throughout the entire project lifecycle, ensuring that potential conflicts or errors are identified and resolved in a timely manner. This collaborative approach ultimately leads to improved quality of the overall design, more efficient project delivery, cost savings and a higher overall quality of the final product.

General maintenance

- Before sharing, ensure that disciplines each exchange detached local copies of the model from their central files.
- Remove all external links and non-design elements, including 3D and 2D design data and floating items.
- Purge the model to eliminate unused or unnecessary elements.
- Ensure that the model is lightweight, error-free and contains the agreed classification data.
- Carefully model each element, such as walls, ceilings, floor finishes and MEP installations, to represent the method of construction and account for MEP maintenance clearance.

- Allocate, name and organise worksets logically, dividing the model into enough worksets to prevent workflow congestion.
- Use placeholder worksets for families belonging to other disciplines.
- Include a title page with the revision number and changes made to the model.
- Ensure that the drawing's title block information meets the project requirements.
- Properly segregate geometry to maintain performance and workability on available hardware.
- Test the Industry Foundation Classes (IFC)-SPF file export to ensure seamless functionality during both export and import processes with other disciplines.
- Ensure that the IFC-SPF is exported to the agreed version, and that the output contains all the relevant data and is formatted according to the terms set out by the delivery team within the BEP.
- A unique disciplinary model is accepted unless agreed differently in the BEP.
- All disciplines must update models based on agreed design changes and ensure that all drawings uploaded to the common data environment are extracted from the federated model, without exception.

Shared coordinates
- Establish a shared coordinate system for the project, ensuring that all models align with it.
- Define true north in the model for accurate orientation.
- Verify the coordination view to make sure that all items to be coordinated are visible and that the section box is appropriately positioned around them.

Modelling best practices
- Model each floor finish separately from structural floor components, ensuring that floors are subdivided by room.
- Model floor finish independently of thickness.
- Represent each ceiling plane as a distinct element and room.
- Include insulation in the model for any ceiling that requires it.
- Ensure that all finishes are incorporated in the wall data, regardless of wall thickness.
- Connect all walls to the top of the slab at the bottom and the bottom of the slab at the top, or provide a clear representation of the intended construction.
- Separate walls by level and take into account the builders' work holes.
- Different worksets should be used for different MEP installations.
- During the modelling process, consider such factors as pipe insulation, hangers, brackets and other accessories that impair ceiling coordination.
- Using appropriate bounding elements, create rooms and spaces for all proposed areas.
- Space should cover from floor to soffit, and all riser spaces must be named in the model.
- All elements within the model should be correctly named, categorised and classified as defined in the BEP.

7.10. Project handover and O&M documents

The importance of O&M is frequently underestimated
In the construction industry, it is commonly observed that operations and maintenance (O&M) documentation is frequently not paid the necessary attention during project execution. This

information is crucial, and all stakeholders, including the design and delivery team, as well as the client's team, play a key role in ensuring its proper delivery.

A common issue is the inaccuracy of project information at the handover stage; this problem is frequently overlooked because of insufficient review by the client's team. As a result, provision of the O&M information is often treated as a mere formality, and the information is often not recognised for its critical value. Its true importance is only realised when clients actively audit and highlight non-compliance issues.

The focus in the construction industry tends to be predominantly on the design-and-build stages, with less appreciation for the accuracy and value of O&M information. However, this information is essential for the client to effectively manage the facility after completion.

The significance of reviewing and verifying project information at various stages, especially during construction, cannot be overstated. This enables the client to assess the quality of the deliverables and to hold the lead appointed party accountable for any discrepancies or omissions. Unfortunately, these revisions often occur too late, either towards the project's end or during the initial aftercare period, making it challenging to obtain updated or missing information from the supply chain.

To ensure accurate and comprehensive project information to support operations and maintenance, it's essential to maintain a 'golden thread of information' throughout the project. The lead appointed party must prioritise this from the beginning and at every project stage. This approach ensures compliance with industry standards and regulations and increases end user satisfaction. Upholding the golden thread of information guarantees the consistency, accuracy and accessibility of critical data for effective operations and maintenance. In this way, clients can be confident that the handover information meets their expectations and enables optimal utilisation of information during the operations and maintenance phase. This approach fosters a culture of accountability, transparency and continuous improvement, highlighting the significance of O&M information in achieving successful project outcomes.

The handover stage

During the handover stage, the contractor formally transfers control of the site to the client. This typically occurs after all necessary inspections and tests have been completed and the practical completion has been certified. At this point, the contractor has the responsibility to provide the client with all necessary documentation and training for the proper operation and maintenance of the facility throughout its lifecycle. It is crucial that this stage is well-planned and executed to ensure a smooth transition from the construction phase to the operational phase. This includes minimising potential issues and disruptions, and ensuring that the facility is ready for use as soon as possible to allow the client to operate the facility effectively and efficiently.

It is often the case that clients do not specify the structure and content of the operations and maintenance information, needed to support facility management, early on in the project. Therefore, it is crucial to involve end users during the design and construction phases to ensure that these specific needs and requirements are met. This is important not only in terms of design and construction but also from the asset information perspective. Actively seeking and incorporating clients' input and collaboration throughout the design and construction processes is essential to ensure their needs are covered and to bridge the gap between the design and construction processes and the efficient operation and maintenance of the facility.

However, it is not only the client who benefits from this approach. The lead appointed party also gains significantly by gaining a clearer understanding of the client's expectations and requirements, as well as a greater level of trust and confidence in the accuracy of the information provided, and benefits from a more streamlined and efficient process. This ensures that there is no need to repeat the same tasks after handover, owing to missing or inaccurate information.

To ensure that the client can properly operate and maintain the facility, it's crucial that the provided information adheres to key principles. There is a range of documentation to be provided, such as maintenance manuals, drawings, warranties and instructions for periodic testing, inspection and repairs, along with essential training for facility personnel.

At the handover stage, the documentation supplied to the appointing party plays a vital role in supporting the operations and maintenance phase. For effective facility management, the information should ideally follow these principles, which are examples and not an exhaustive list.

- *Accuracy.* Documents must be up to date, validated and verified by the lead appointed party, reflecting the work executed during construction.
- *Accessibility.* Documentation should be available in an electronic format, making it easy to access, navigate and retrieve information.
- *Completeness.* The documentation should include all relevant details, such as the latest construction issue drawings, O&M manuals, warranties and any other necessary documentation requested by the appointing party.
- *Clarity.* Information should be presented in a straightforward manner, with clear explanations and illustrations where appropriate, to aid the reader in understanding the content.
- *Relevance.* It's essential to guarantee that all provided information is directly pertinent and useful for the facility's operation and maintenance. This means avoiding the use of generic templates that are not specifically tailored to the project.

As mentioned earlier in this book, there are many instances in which we work on projects where we need information about the existing building conditions. However, commonly, one or more of these requirements are not met. As a result, our design and construction process cannot be supported by existing information, even if it is available, and we must conduct surveys to capture the correct information about the building. This leads, before construction can begin, to initial costs and time that must be considered when planning the works at the tender stage.

The end user of a facility might face several challenges during its operation if the O&M documentation is incorrect or inadequate. The following issues might arise.

- *Decision-making challenges.* Without comprehensive information that adheres to principles of accuracy, accessibility, completeness and clarity, facility managers might struggle to make informed decisions regarding the operation and maintenance of the facility. This can lead to operational inefficiencies, potential equipment failures and additional costs over the asset's lifecycle.
- *Clarity and accuracy.* If the O&M documentation lacks clarity or contains inaccuracies, or if information is misplaced or missing, this can lead to operational challenges and disruptions. These issues can affect the facility management team's efficiency and result in operational errors.
- *Problems with warranty claims.* O&M documents often include warranties for equipment and systems. Inaccurate or incomplete documentation can make it difficult for the facility

management team to locate the necessary documentation to claim these warranties, resulting in additional avoidable expenses.

- *Cost escalation risks.* Operating and maintaining a facility requires specific knowledge and expertise. If the O&M documentation is not clear, complete, relevant, accurate or easily accessible, the client might have to hire external contractors for essential surveys and maintenance tasks – costs that could be avoided with proper documentation.

These potential problems underline the importance of accurate and comprehensive O&M documentation for the successful handover and operation of a facility. To ensure smooth facility operation and to avoid extra costs and challenges, O&M documentation for construction projects should be meticulously prepared and reviewed. Adopting a proactive approach will not only mitigate operational problems but will also expedite the handover of the facility to the client.

What makes delivering correct O&M so challenging?

The construction industry has a history of producing O&M manuals that are of poor quality and not delivered in a timely manner, despite the benefits that this information provides to the end user. The process of producing accurate O&M information is a complex task that requires coordination and active collaboration from all parties involved in the project, including the end user. This can be particularly challenging when the project is large and complex.

Each project has its own set of requirements and specifications, which cannot always be generalised or standardised. Because this industry comprises a wide variety of supply chains, gathering uniform datasets and formats from various subcontractors becomes an uphill task. It's common for subcontractors to use preset templates for data collection that don't always correspond to the on-site work's specificities. Such practices can inadvertently contribute to discrepancies in the data collected. This variance can ultimately reduce the quality and accuracy of the O&M manuals that are produced.

Despite attempts to improve the quality of client project data by following international best practices for product data creation and management, there has been limited success. The emphasis has been on standardising the production, usage and management of data through collaborative product data templates and a software platform. However, as of the time of writing, substantial progress or real-world applications in this area are yet to be observed.

Additionally, as identified at the beginning of this section, the O&M aspects often do not receive the attention they deserve. It is common practice to leave the creation of O&M manuals until the end of a construction project. The primary focus is often on completing the manuals quickly to meet project completion deadlines, instead of collaborating with the client throughout the project's duration. This rush often results in the main contractor hastily going through the process of creating O&M manuals, viewing them as a necessary yet burdensome requirement to achieve practical completion. Contractors often overlook their significance to the client and the overall project.

As a result, subcontractors frequently resort to hiring external consultants to hurriedly compile all the documents received from suppliers and subcontractors in a single file, without emphasising the quality of the final product.

Furthermore, it is a common issue in many organisations that the verification process for O&M is inadequate. This process is not always carried out by senior staff or appropriate team members, owing to time constraints, and is instead conducted by others, who might not have the necessary expertise and knowledge to validate and verify the accuracy of the content provided by the different parties involved. Also, client involvement in the creation of the O&M manual is often delayed and not carried out in conjunction with the asset facilities team, who typically become involved only after the construction project is completed.

These factors contribute to the production of O&M manuals that are often chaotic and of poor quality. The lack of adequate planning, availability and accountability of the teams involved in creating the manuals, as well as the performance of external consultants and the systems used to produce them, all play a major role in this issue.

Common issues within O&M manuals often include the following.

- *Stakeholder engagement deficit.* A lack of input from clients and facility managers often leads to the production of manuals that fail to address project-specific challenges and overlook practical, operational requirements. This absence of collaboration results in manuals not adhering to principles of being accurate, accessible, complete, clear and relevant.
- *Content quality and relevance issues.* Manuals compiled from generic manufacturers' information or created hastily at the project's end tend to contain outdated, inaccurate or incomplete information. This approach neglects asset-specific or project-specific needs and fails to meet legal requirements effectively.
- *Usability and training shortfalls.* Complex language filled with industry jargon, coupled with unintuitive interfaces, makes manuals difficult for non-experts to understand. Additionally, insufficient training in manual usage leads to ineffective operation and maintenance of assets.
- *Structured handover and documentation practices.* The lack of a structured handover process and the omission of client and facility management team involvement in manual preparation contribute to information transfer gaps. Manuals prepared in this manner are unlikely to support effective operations and maintenance, owing to poor usability and a lack of detailed, relevant content.

What information should be provided as part of the O&M documentation?

Being part of the overall asset information model, an O&M manual is a crucial document for any project as it contains all the necessary information needed to operate and maintain the facility once the construction is completed. It is important to note that each project will have unique requirements, depending on the type of project and work involved.

Therefore, it is essential to involve the end user and the facility management team from the beginning of the project to define the appropriate level of information needed.

The following are some typical contents of an O&M manual.

- *Updated drawings.* The latest architectural and equipment drawings, formatted as agreed, showing where everything is located and how systems are connected.
- *Asset register.* A list and description of all systems and equipment in the facility.
- *Completion certificates.* Official documents confirming that construction phases or equipment installations have been completed correctly.
- *Project overview.* Key information about the project itself, the work carried out and the suppliers involved.
- *Original specifications and changes.* A list of the original requirements for the project, along with any changes made during construction.

- *User guidelines.* Comprehensive instructions for operating and maintaining the facility. This should include details on equipment and systems, as well as any manufacturer guidelines and warranties.

- *Test certificates.* Documents that confirm successful commissioning and testing, including fire, gas and electrical safety certificates.

- *Equipment warranties.* Information on warranties for systems and equipment, including how to make a claim and any spare parts supplied.

- *Safety procedures.* Health and safety guidelines and emergency protocols, considering any remaining hazards in the facility.

- *Environmental assessments.* Information on any assessments conducted to measure the environmental impact of the facility, along with corresponding certificates.

- *Training materials.* Resources to help train the client's staff in the safe and efficient operation and maintenance of the facility.

- *Maintenance plans.* A schedule outlining when and how to carry out regular service and maintenance tasks.

In addition, any other information that has been agreed on as part of the project deliverables should be included. This includes models in the agreed formats and structured data in the specified format, such as COBie.

Actions to be considered

Before the production of O&M documentation
To ensure the smooth and efficient handover of a construction project, it is crucial for the client to have a clear plan from the outset. This plan should guide task teams in organising and compiling all necessary documents, with the aim of minimising errors and incomplete records.

Here are some important steps to include in the plan.

- *Comprehensive strategy.* Define all requirements and responsibilities for each team member, covering deliverables, milestones and specific tasks.

- *Communication.* Effective communication is key to avoiding misunderstandings and ensuring that all parties meet expectations. Clearly inform all parties about the production of required O&M documents and the handover process, making sure that everyone understands their roles and responsibilities for an efficient handover.

- *Information transfer.* Early in the project, establish a clear strategy for the transfer and storage of all project information, as required by the client. This will allow easy access for the client's facility management team.

During the production of O&M documentation
The creation of effective O&M manuals requires specific actions during the handover phase.

- *Assign a representative.* Each task team should designate an individual to take responsibility for gathering and organising necessary content and collaborating with the lead appointed party to meet deadlines. This representative needs to be knowledgeable and have access to the necessary resources. Overloading temporary administrators or site teams with the assembly and validation of technical content is not practical.

- *Early training.* Provide all involved parties with training on the information management solution used to gather and manage information. This approach ensures timely information provision and reduces the need for extensive validation of O&M information at the project's end.

- *Tailor process and tools.* Adapt the process and tools for managing O&M information to the client's specific needs, ensuring that data are easily accessible and the system is user-friendly.

- *Ensure accuracy and comprehension.* The lead appointed party is responsible for providing accurate information regarding the work completed, according to approved design specifications. The lead appointed party should also be aware of the potential contractual consequences of providing inaccurate information.

- *Regular progress monitoring.* Conduct frequent progress reviews and updates. Ensure that tasks follow the plan, addressing any issues promptly to maintain project momentum and achieve a smooth handover.

- *Validation and verification.* The lead appointed party must establish a clear process and define accountabilities for the validation and verification of project information.

- *Continuous collection of information.* Maintain an ongoing collection of O&M information throughout the project. Avoid leaving the compilation of all data until the end, as this can lead to inefficiencies and information gaps.

After the production of O&M documentation
After handover, maintaining a support system for the client is critical.

- *Ongoing support.* Continuously assist the client in addressing any queries or concerns related to the operation and maintenance of the facility.

- *Feedback.* Request feedback to gauge the effectiveness of the handover process and the usability of the O&M information. This feedback, as part of the learning culture within the business, can be utilised for continuous improvement in future projects.

- *Regular updates.* Ensure that any changes in the facility are accurately reflected in the O&M manuals so that they truly represent the project. Regular revision of the documentation is important to keep it current and useful for the client's ongoing facility management needs.

Table 7.2 lists items to consider before, during and after the production of O&M documents, to improve the delivery of accurate information to the appointing party.

7.11. Asset information model

What is asset management?
Asset management, guided by ISO 55000 (BSI, 2014), aims to align a company's activities with its long-term goals to maximise asset value (Institute of Asset Management, 2014). Unlike the past, where asset decisions were often isolated, today's approach is unified and data-driven. Everyone, from top management to the different task teams, collaborates to make informed decisions.

Reliable data are crucial for good decision making, and companies now use a single, dependable source of information. Collaboration across departments is key, often requiring a culture shift for best results.

The Institute of Asset Management (IAM) provides guidance for effective asset management, boosting confidence in investors and leaders. However, challenges still exist, for example, departmental silos, which can result in poor decisions. Change can come from either grassroots employee initiatives or top-down strategies, often driven by external factors. The aim is to create a culture of continuous learning and improvement, leading to greater innovation and fewer problems.

Table 7.2 Checklist: operations and maintenance considerations

Before production

Comprehensive strategy

- ☑ Define requirements for each team member.
- ☑ Outline deliverables, milestones and specific tasks.

Communication

- ☑ Inform all parties about the production of required O&M documentation and the handover process.
- ☑ Clarify roles and responsibilities.

Information transfer

- ☑ Establish a strategy for transferring and storing project information.
- ☑ Ensure easy access for the client's facility management team.

During production

Assign a representative

- ☑ Designate an individual for each task team.
- ☑ Ensure that the representative has the necessary knowledge and resources.

Early training

- ☑ Train parties in using the information management solution.
- ☑ Aim for timely information provision.

Tailor process and tools

- ☑ Adapt tools and processes to the client's specific needs.
- ☑ Ensure data accessibility and user-friendliness.

Ensure accuracy and comprehension

- ☑ Provide accurate information, in accordance with design specifications.
- ☑ Be aware of potential contractual consequences.

Regular progress monitoring

- ☑ Conduct frequent reviews and updates.
- ☑ Address issues promptly.

Validation and verification

- ☑ Establish a clear process for validation and verification.
- ☑ Define accountabilities.

Continuous collection of information

- ☑ Maintain ongoing data collection.
- ☑ Avoid last-minute compilations.

After production

Ongoing support

- ☑ Assist the client with queries or concerns.

Feedback

- ☑ Request feedback on handover process and the quality of the information provided.
- ☑ Utilise feedback for continuous improvement.

Regular updates

- ☑ Update O&M manuals to reflect facility changes.
- ☑ Keep documentation current and useful.

What is the asset information model?

The asset information model (AIM), as outlined in ISO 19650-1 (BSI, 2019), serves as a centralised repository for both strategic planning and daily asset management activities set by the appointing party. The AIM is not just a collection of static documents; it's a dynamic resource that evolves throughout the asset's lifecycle. It includes information crucial for initiating new projects, such as equipment registers, maintenance costs and installation schedules, along with such details as property ownership.

When establishing BIM requirements for a project, the appointing party should ensure that these align with broader asset management objectives. The AIM thus becomes a single source of truth that supports effective decision making across various activities.

The AIM should comprehensively encompass managerial, legal, technical, financial and commercial information. This is important for making well-informed decisions, be they related to maintenance, renovations or legal compliance. Such data should also align with the exchange information requirements and be systematically integrated in the project's various phases. For further guidance, ISO 19650-3 (BSI, 2020a) provides additional details on the operational phase of the assets. By adhering to these standards, the AIM ensures that all stakeholders operate from a unified, authoritative data source, thereby enhancing the effectiveness and efficiency of asset management efforts.

The ISO 19650-3 (BSI, 2020a) includes a comprehensive overview of the asset lifecycle. Figure 7.8, which is based on this standard, encompasses both the operational and delivery phases.

Figure 7.8 Generic project and asset information management lifecycle (adapted from BSI, 2020a)

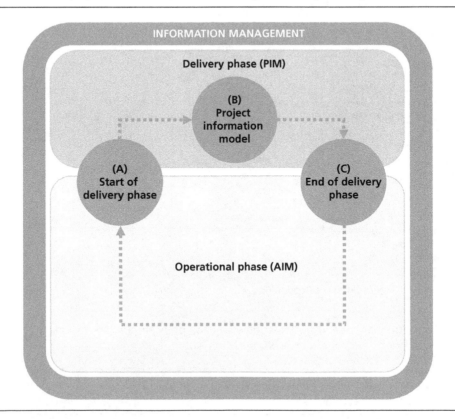

As shown in Figure 7.8, the delivery phase is divided into three key stages.

- *Stage A.* This marks the beginning and involves transferring essential data from the AIM to the project information model (PIM).
- *Stage B.* At this juncture, there's a continuous evolution of the initial design model, transforming it into the virtual construction model.
- *Stage C.* This marks the end of the delivery phase, with the appropriate vital information moving from the PIM back to the AIM.

The AIM can encapsulate both graphical and non-graphical data, as well as documents and meta-data. It can be associated with a single asset or a portfolio of assets. However, traditionally, the AIM is often inadequately maintained throughout its lifecycle. This lack of maintenance leads to an insufficient supply of graphical and non-graphical information from the appointing party to support further works on the existing facility. As a result, it necessitates new surveys and verifications, as the information is not sufficiently reliable to facilitate the design development.

To curb costs at the outset of the delivery phase when transitioning from AIM to PIM, maintaining an up-to-date AIM is crucial. An updated AIM not only assists the client in minimising costs but also ensures the availability of reliable information for the development of the design. This approach underscores the importance of the AIM, underpinning its role in strategic and operational asset management while also providing a robust foundation for future project initiation.

How to define and maintain the AIM

Defining the project deliverables that need to be included in the AIM at the project's outset, as part of the exchange information requirements, is a main action for the appointing party. This ensures that all necessary information is meticulously captured and organised in a structured format, allowing for efficient use and streamlined management of the asset throughout its lifecycle.

The AIM is indispensable for organisations looking to optimise their asset management. Yet there is a prevalent ambiguity among many clients about the kind of information necessary for the operations and maintenance phase of their assets. This often results in inefficient management of the asset during its lifecycle.

As previously highlighted, for effective asset management, organisations must have a robust strategy in place, encompassing clear organisational information requirements (OIR) and asset information requirements (AIR). Both OIR and AIR are pivotal, granting organisations a lucid perspective of the requisite information for managing varied assets.

Organisations should consider the following strategies to meticulously define, optimise, secure and effectively deliver their asset information model (AIM), ensuring they serve both operational targets and the end users' needs throughout the asset's lifecycle.

- Specify the necessary information to be included in both the structured asset data and the O&M documentation.
- Engage the facility management team and end users in producing project documentation. Agree a strategy for data collection at various phases to ensure necessary and relevant structured data capture tailored to the project.
- During the asset's design and construction, actively participate to meet operational targets, pinpoint improvements and ensure that the asset meets end users' needs.

- Adopt a security-minded approach with robust data management policies, considering legal storage requirements, backups and disaster recovery.
- Consider requirements to make the asset management information easily accessible and user-friendly to support its accessibility and maintenance.
- Set clear policies for managing access to information throughout the asset's lifecycle.
- Evaluate the suitability of current systems for the effective handling and maintenance of the AIM.

After defining the asset information requirements, it's crucial to establish the roles and responsibilities of team members throughout the asset's lifecycle. This ensures that data collection, verification, validation and maintenance meet the requirements. Appointing a responsible party to oversee these tasks is essential to make sure that the AIM effectively supports decision making, safety and regulatory compliance.

During the execution of the project, it's important that data are collected progressively and assessed at each stage, rather than towards the end. Adopting a progressive data collection approach throughout the project's duration is best practice. Collecting, validating and verifying data in increments allows for the early detection and rectification of discrepancies, helping to circumvent potential extensive downstream issues. This approach also ensures that the data adhere to the EIR's specified format and that the information sourced from the task teams is precise, thereby effectively supporting the client's needs.

Once the project is complete, maintaining the AIM during the operational phase is necessary. Regular updates are needed, as the data can quickly become obsolete after the handover stage. Having a system to track changes and updates is also required. A maintenance plan, including a schedule for regular updates and a process for addressing issues, should be developed.

A well-maintained AIM provides organisations with a foundation for effective asset management, enabling better decision making, improved efficiency, enhanced safety and regulatory compliance.

Organisations should consider the following to maintain the AIM.

- Provide ongoing training for employees to ensure that they can access and update the AIM effectively. Engage with stakeholders for continuous improvement, incorporating a feedback loop for suggestions and advanced learning.
- Assign tasks within the team for regular reviews of the AIM throughout the operational phase. Ensure that the model reflects the asset's current state, and any modifications, and remains compliant with the AIR. Implement routine audits and adhere to industry standards and regulations.
- Secure a commitment from the design and construction team during the aftercare period to address necessary actions promptly and ensure continuous alignment with project goals.
- Adhere to a security-minded approach and manage access controls, periodically reviewing and revoking access as needed.
- Update the information to keep pace with technological advances, ensuring user accessibility. Integrate new technologies and establish clear procedures for managing updates to the AIM.
- Record and share lessons learned from previous projects to improve how assets are managed in the future and avoid making the same mistakes. Track the effectiveness of these practices through performance metrics.
- Archive any information that's no longer relevant or trustworthy and include disaster recovery and backup plans to maintain data integrity.

Challenges

In my experience as a main contractor, delivering asset data encompasses navigating several potential challenges.

▪ *Unclear initial requirements.* A lack of clear appointing party requirements at the start of the project might lead to deliverables not meeting expectations regarding format and content. A thorough communication and understanding of the appointing party's needs is crucial to avoid this issue.

▪ *Poor-quality data.* The data provided by each task team must be accurate and complete, according to the appointing party's requirements. All stakeholders need to be accountable for providing high-quality information, and the appointed lead party should validate and verify the information.

▪ *Limited FM team engagement.* A lack of facility management (FM) team involvement or insufficient feedback during the project could cause a misalignment with expectations. Early and consistent engagement with the FM team and appointing party is essential.

▪ *Delayed information delivery.* Consistent information delivery throughout the project is crucial, as waiting until the end might affect the AIM's quality. You might face challenges in obtaining information from the task teams. These challenges need to be mitigated and managed during the appointment.

▪ *Asset data consolidation.* When handling existing buildings, merging asset data between new and existing assets is crucial. Checking the accuracy of existing information and aligning it with the new is essential.

▪ *Validating existing assets.* If there's no reliable data source for existing assets, measures to validate any retained services that need to be part of the AIM should be considered from the project's outset.

▪ *Data delivery discrepancies.* The EIR might demand data using COBie, but the FM team might have different requirements. Early engagement can prevent such conflicts and ensure that the information is requested in the correct format and structure.

▪ *Data inconsistencies.* Discrepancies between data sources from task teams (such as differing pipe sizes in the model and schematic drawing) can lead to incorrect and contradictory information. Extracting data from a single source is essential, and the task teams and lead appointed party must ensure data accuracy.

▪ *Sensitive information handling.* The AIM might contain sensitive data that must be protected. Limiting access to authorised personnel is paramount. However, these measures must be well-defined and not excessive to avoid creating burdensome processes and distractions, which could hinder the efficient completion of the project.

▪ *Maintenance responsibilities.* The appointing party should clearly define roles for maintenance and ensure that there is a designated role responsible for liaising with the lead appointed party to address any inconsistencies during the project and at the handover stage. This role should also include the ability to update the AIM during the operations and maintenance stages, to prevent it from becoming outdated after handover.

Benefits

While I have covered the benefits at the asset level in Section 5.8, PwC (2020) produced a report titled *Assessment of the Benefits of BIM in Asset Management. Part 1: Context and Methodology*, which describes how the benefits of BIM in asset management can be realised at three distinct levels.

- Asset level
    - □ *Quick access.* Using a digital AIM offers faster and more convenient access to crucial data than traditional paper manuals.
    - □ *Improved planning.* The easy access to reliable information through the AIM boosts the maintenance team's confidence, facilitates better planning and can speed up the execution of tasks.
    - □ *Reduced site visits.* The AIM eliminates the need for numerous site visits caused by unexpected conditions, as it provides all the necessary data upfront. This results in time savings for the asset manager.
- Portfolio of assets level
    - □ *Asset understanding.* Utilising different AIMs for a collection of buildings provides a clearer view of asset conditions. This promotes efficient and expedited maintenance work.
    - □ *Time efficiency.* By carefully planning the sequence of maintenance tasks across the building portfolio, travel time is minimised. This leads to a quicker overall maintenance process, saving valuable time.
- Business level, where many portfolios of assets are managed
    - □ *Accurate planning.* Leveraging BIM for asset data aids in precise maintenance planning for a number of buildings. This includes calculating the demand for materials and equipment accurately.
    - □ *Smart procurement.* Implementing informed procurement strategies, such as bulk buying or negotiating better supplier contracts, offers economies of scale. This is more cost-effective compared with individual building purchases.
    - □ *Cost savings.* Adopting an intelligent approach to procurement leads to significant reductions in maintenance costs across the business. This is advantageous for the asset owner.

These examples illustrate a structured, multilevel approach to realising and aggregating benefits from asset information management, with each level building on the efficiencies and insights gained from the previous one, leading to time and cost savings in building maintenance operations.

Additionally, the content of the report expands on existing and potential new impact pathways, with examples of activities at the asset, portfolio and business levels.

- Strategy
    - □ *Strategic planning.* Having access to asset data can significantly aid in decision making and planning at both the portfolio and business levels, especially in terms of delivering and managing assets effectively.
- Design, handover and commissioning, and operation and end of life
    - □ *Financial management.* Utilising BIM-enabled asset data saves time and enhances the accuracy of forecasting and managing asset management costs.
    - □ *Maintenance.* Maintenance processes can be improved by leveraging digital asset information to devise precise maintenance and repair plans.
    - □ *Facility management.* By utilising asset information, operations and maintenance of asset components related to security and surveillance can be optimised and coordinated with janitorial services.
    - □ *Inspection and monitoring.* Digitalising asset data, such as information on component deterioration, facilitates predictive maintenance planning across different assets.

- ☐ *Compliance management.* Meeting regulatory or contractual reporting requirements on assets is expedited and made more accurate with the use of digital asset datasets and tools.
- ☐ *H&S management.* Digital asset information aids in health and safety (H&S) training for specific operational and maintenance activities – for instance, using 3D models to acquaint staff with hazards before site visits.
- ☐ *Risk management.* Asset information is crucial to identify and assess risks where an asset or its components might be exposed to natural hazards, extreme weather conditions or fire, aiding in the formulation of emergency response plans.
- ☐ *People management.* Based on staff suitability and asset information, the appropriate personnel can be identified to conduct specific asset management activities.
- ☐ *Sustainability and environmental management.* Based on BIM-enabled energy efficiency information, an asset's energy consumption can be assessed and energy saving opportunities can be exploited.
- ☐ *Quality management.* Digital asset information can be aligned with relevant specifications and standards, enhancing quality management throughout an asset's lifecycle. Furthermore, this information can undergo data quality diagnostics and remediation to ensure ongoing data quality improvement.

In summary, the PwC (2020) report highlights the significant benefits of using BIM in asset management. These benefits span from improved efficiency and planning at the asset level to cost savings and smart procurement strategies at the business level. This approach not only enhances operational efficiency but also leads to considerable cost reductions.

7.12. Golden thread

The golden thread, which is a key concept in the Building Safety Act 2022 (Department for Levelling Up, Housing and Communities, 2022; HMG, 2022), provides an opportunity to ensure robust information management practices for the future of our industry. It needs to be part of any comprehensive information management strategy within a business; therefore, raising awareness about these principles is essential.

The maturity of the Building Safety Act 2022 requires constant attention. As its full implementation was scheduled for October 2023, it's necessary to thoroughly understand these principles. By this date, building owners were expected to have their building safety measures fully in place.

The UK Government selected Dame Judith Hackitt to conduct an impartial review of building regulations and fire safety following the Grenfell Tower disaster. Her report, *Building a Safer Future* (Hackitt, 2018), suggested the implementation of a 'golden thread'. The idea behind this concept is that buildings are treated as comprehensive systems, and critical information for robust safety management is supplied throughout the building lifecycle.

The Government has pledged to implement this suggestion. Consequently, the golden thread will be relevant to all structures under the new building safety regime. Initiated through the Building Safety Act 2022 (HMG, 2022), this regime introduces a significant opportunity to transform information management, placing responsibility for project safety information on the client and extending accountability to all project stakeholders.

In this context, the golden thread becomes a vital component of your business's information management strategy. It offers a structured approach to store building information in a manner that supports dynamic data sharing, facilitates easy tracking and establishes a consistent and reliable source of truth. Moreover, it functions as tangible proof of risk mitigation efforts, going beyond mere statements, demonstrating proactive safety management.

This measure is aimed at rectifying the industry's issues concerning the efficient management and availability of key information throughout a building's lifecycle. In the past, there has been a lack of vital data for relevant individuals to ensure the safety of their buildings.

The definition of 'golden thread', as stated by the Building Regulations Advisory Committee (Davies *et al.*, 2021), consists of seven key aspects.

- The golden thread is a collection of critical information for those responsible for a building, evidencing compliance with regulations during construction and across the building's lifecycle. It helps manage and mitigate building safety risks, including fire or structural threats.
- The information stored in the golden thread is regularly reviewed and updated to maintain its relevance and usefulness.
- The golden thread includes both the information and documents and the information management processes to support building safety.
- The golden thread information should be digitally stored in a structured way, following Government-issued digital standards and principles.
- The golden thread approach applies throughout a building's lifecycle and supports a culture of building safety.
- Building safety includes fire and structural safety, as well as the safety of all people in and around a building.
- Various stakeholders, including building managers, architects and subcontractors, will access the golden thread throughout the building's lifecycle. The accountable person is also responsible for sharing this information with relevant parties – for instance, residents and emergency responders.

In essence, the golden thread serves a dual purpose. First, it is the vital information that helps us comprehend a building's structure and function. Second, it outlines the necessary procedures to maintain the safety of both the building and its occupants, both in the present and moving forward.

The BRAC working group (Davies *et al.*, 2021) has developed a set of golden thread principles to support the definition of the golden thread, which can be summarised as follows.

- The golden thread must be accurate, trusted and maintained for building safety and compliance with building regulations. It will be used as evidence to manage building safety risks, and the information should be updated and verified using a clear change control process.
- The golden thread will provide residents with accurate information about their homes and hold accountable persons and building safety managers responsible for building safety. A properly maintained golden thread will assure residents that their building is being managed safely.
- The golden thread will support culture change in the industry, leading to increased competence, updated processes and a focus on information management and control.

- The golden thread will bring all information together in a single place, reducing duplication of information and improving accountability and responsibility.

- The golden thread must be securely managed, adhering to both the General Data Protection Regulation (GDPR) (EC, 2016) and the Data Protection Act 2018 (HMG, 2018), to ensure the protection of personal information and maintain the security of the building and its residents.

- The golden thread will record changes and who made them to improve accountability. There is accountability at every level, from the clients or the accountable person to those designing, building or maintaining a building.

- The golden thread must be clear, understandable and focused on the needs of the user, using standard methods and consistent terminology, so that it can be used effectively.

- The golden thread must be accessible, stored in a structured way and presented in such a way that people can easily find, update and extract the right information.

- The golden thread must be formatted in a way that can be handed over and maintained over the entire lifetime of a building, conforming to open data rules and principles of interoperability.

- The golden thread should only keep relevant and useful information for building safety, and it should be periodically reviewed to ensure that this is the case.

The golden thread and the ISO 19650 series (BSI, 2019, 2020a, 2020b, 2021, 2022a) are closely related, as they both place an emphasis on efficient and effective management of building information. Correct implementation of the principles outlined in ISO 19650 will support the implementation of the golden thread. By adopting the ISO 19650 standards, an organisation can establish a framework that meets the golden thread requirements. This involves ensuring that information is accurate, accessible and up to date, which is essential for building safety and ongoing management.

The significant repercussions of non-compliance with the Building Safety Act 2022 (HMG, 2022) underscore the necessity for companies to diligently manage project information. This is no longer an optional best practice, but an essential duty to ensure safety and regulatory compliance. Failure to comply carries severe consequences, ranging from endangering public safety to incurring severe fines and legal action, potentially including up to 2 years of imprisonment for the accountable parties.

Regrettably, it's often the threat of substantial fines that prompts companies to manage their information properly. Ideally, this should be an innate part of their operational strategy, cultivated through the right culture and behaviours.

Good information management not only ensures regulatory compliance but also offers numerous advantages for project execution and overall business performance. Therefore, it's crucial for companies to shift perspective and view information management not just as a means to avoid penalties, but as a proactive approach that contributes to safety, efficiency and success in projects and overall businesses.

Building Safety Act 2022 (HMG, 2022)

What it is and why does it matter

This Act marks a crucial turning point for the construction industry. It's set to reshape our sector, our business practices and the way we all will work going forward. The way in which buildings are designed, built and maintained will see a fundamental change, including the records of materials used, who installed them and how they were maintained.

The Building Safety Act 2022 (HMG, 2022) is broad in scope, covering every aspect of building safety, not just fire safety, and is applicable to all buildings. However, it gives special focus to the processes involved in building or refurbishing higher-risk buildings.

The Act requires detailed record-keeping on material usage, installation and maintenance, bringing the construction industry up to par with, for example, the nuclear and aviation sectors, in terms of regulatory compliance. It introduces new regulatory bodies and extends liability for non-compliant work, ensuring a higher standard of safety and accountability.

Moreover, the Act enforces strict penalties and fines, similar to existing health and safety (H&S) laws; H&S laws protect individuals during construction, while the Building Safety Act 2022 (HMG, 2022) safeguards occupants in the long term. It is essential to pay ongoing attention to the Act, as it will continue to develop, with full implementation set for April 2024, necessitating a thorough understanding of its principles.

Some considerations of the Act

- *Enforcement tools.* Among the new measures, the Act empowers the Building Safety Regulator with new enforcement tools, such as issuing stop and compliance notices that can result in unlimited fines and up to 2 years' imprisonment, extending the remedial action period to 10 years, and enabling individuals to file civil claims for historical non-compliance with building regulations.
- *Regulatory bodies.* Three regulatory bodies are established to handle complaints, each with defined roles and responsibilities: the Building Safety Regulator, the Regulator for Construction Products and the New Homes Ombudsman Service.
- *Responsibility definitions.* The Act defines responsibilities for accountable persons as building owners or operators, assigns liabilities to building owners for existing structures and sets compliance obligations for duty holders, including the client, principal designer, principal contractor and other designers and contractors. All parties will need to demonstrate competence in their respective areas.
- *Golden thread of information.* A continuous information thread, extending from project inception to ongoing maintenance, is introduced.
- *Higher-risk buildings (HRBs).* A three-gateway approval process is identified for HRBs, based on specified criteria, including height, use and occupancy.
- *Legislative amendments.* Changes to existing legislation – for example, extending liability under the Defective Premises Act 1972 (HMG, 1972) to 15 years for new residential buildings.

MAIN CONTRACTOR'S ACTIONS

The actions that need to be considered include the following.

- *Adhere to stringent regulations.* Comply with enhanced regulations to ensure accountability in fire and structural safety.
- *Ensure worker competence.* Guarantee that all workers and subcontractors are skilled and qualified, thus enhancing overall building safety.
- *Maintain the golden thread of information.* Manage, record and update construction information, contributing to the golden thread.

- *Navigate gateway stages effectively.* Understand and adhere to the updated project timelines for gateway progression, and obtain necessary approvals at each stage.
- *Implement stringent management of change control.* Rigorously manage changes to the project to ensure compliance and maintain the integrity of the golden thread.
- *Prepare and submit necessary documentation.* Comprehend, manage and submit all required new documentation.

CLIENT'S ACTIONS

The actions that need to be considered include the following.

- *Verify and appoint key roles.* Clients should verify the qualifications of the principal designer and contractor and appoint them accordingly.
- *Ensure collaboration and information sharing.* It is crucial for clients to guarantee effective collaboration and information sharing among all project stakeholders.
- *Allocate resources for compliance.* Adequate time and resources must be allocated to adhere to regulatory standards.
- *Appoint accountable persons.* Clients are required to determine and appoint a principal accountable person (PAP).
- *Confirm regulatory adherence.* A written statement confirming adherence to building regulations is necessary.
- *Adhere to the gateway regime.* Clients need to handle submissions for gateway applications and manage any change requests.
- *Assume responsibility for unappointed roles.* Clients must accept responsibility and liability for any mandatory roles that remain unfilled.
- *Provide necessary registration information.* Essential details must be provided to the PAP for building registration prior to occupation.
- *Maintain the golden thread of information.* In compliance with the golden thread of information, clients must ensure that a comprehensive record of building safety information is maintained and accessible.

While I am not an expert on the Act, I recognise that there is much for us all to learn, yet the benefits it offers are readily apparent. It's worth noting that BIM4Housing (2023) is playing a significant role in facilitating communication and understanding of the BSA, greatly contributing to our collective knowledge and implementation efforts.

The Act marks a significant progression in establishing robust legislation for information management, enhancing transparency and accountability. It conforms to ISO 19650 (BSI, 2019, 2020a, 2020b, 2021, 2022a) standards, providing support to professionals who have faced challenges in managing information within their businesses over the years. For those committed to proper information management, the Act is particularly beneficial, as it not only facilitates the adoption of these standards but also underlines the severe consequences of failing to comply.

The implications of the Act are far-reaching, with the Act poised to significantly improve the construction industry overall. This includes bolstering the industry's reputation, safety, quality and accountability, resulting in safeguarding the well-being of end users.

BS 8644-1:2022

On 31 July 2022, British Standard BS 8644-1:2022 (BSI, 2022b) was published in response to the Hackitt Review's recommendations (Hackitt, 2018). This standard provides guidelines on handling fire safety information digitally throughout a built asset's lifecycle. It aims to enhance the safety of the built environment, reduce fire-related disruptions and ensure that the right people have access to the correct fire safety information. The standard also seeks to facilitate better identification of fire safety hazards, clearer allocation of safety responsibilities and wider dissemination of fire safety knowledge. It is important to mention that the standard is designed to be read together with the ISO 19650 (BSI, 2019, 2020a, 2020b, 2021, 2022a) series.

However, an open letter (Stanton, 2022) has been issued, calling for urgent and significant revisions to BS 8644-1 (BSI, 2022b). This letter is signed by professionals in the digital construction industry. It addresses various concerns that have been raised about how this standard is applied in practice. While those working in information management support the ideas behind this standard, they've highlighted four key issues: the prescriptive nature of Section 5, the absence of a comprehensive digital fire safety data exchange schema (FIREie), potential overlap with existing schemas, such as COBie, and ambiguity regarding the path to compliance.

Specifically, the limited amount of technical information in BS 8644-1: (BSI, 2022b) for FIREie could lead to varied interpretations and applications, potentially escalating project costs. While FIREie resembles COBie, the lack of detailed technical information on its use, and the use of both FIREie and COBie schemas together, could result in data redundancy. Additionally, the necessity for exchange requirements focused on a singular purpose, in this case, fire, is also questioned, given the potential need for information exchanges for various other purposes.

Therefore, it's crucial that when the need to implement these practices and guidelines arises, you should communicate clearly with the appointing party to confirm the project requirements and expectations during the tender stage, to avoid unnecessary replication of processes and potential extra expense.

7.13. Digital twins

It's important to provide clarity on the term 'digital twin'. The digital twin is a relatively recent term that has become increasingly popular, yet there is a tendency to misuse or misinterpret it. As someone who is not an expert and has worked on only one project involving digital twins, I feel the need to explore its meaning, especially given the abundance of misleading information on social media about what digital twins are and how they are used. This trend reminds me of the marketing buzz around BIM in the past, where the term was misused, with the wrong definition.

The definition of *digital twin* provided by the Digital Twin Consortium (2020) is:

A digital twin is a virtual representation of real-world entities and processes synchronized at a specified frequency and fidelity.

- Digital twin systems transform business by accelerating holistic understanding, optimal decision making and effective action.
- Digital twins use real-time and historical data to represent the past and present and simulate predicted futures.
- Digital twins are motivated by outcomes, tailored to use cases, powered by integration, built on data, guided by domain knowledge, and implemented in IT/OT systems.

The report *Digital Twin Toolkit: Developing the Business Case for Your Digital Twin* (Abisogun *et al.*, 2021), which was released in February 2021 by the National Digital Twin programme and the Gemini supporters, acknowledges, 'Digital twins still mean different things to different people,' a sentiment that echoes what the industry had said about BIM for many years. Nonetheless, the toolkit identifies a shared characteristic of digital twins: the data as the foundation.

For detailed information on the promotion of digital twin technology through collaboration and knowledge sharing, you can visit the websites of the Digital Twin Consortium (2023) and the Digital Twin Hub (2023). These entities are international and UK-based organisations, respectively, that work towards advancing the development and adoption of digital twin technology.

While digital twins hold great promise for the industry, they are not without their challenges. As for the situation with BIM, digital twins have yet to acquire a universally accepted definition. As pointed out, there are still differing interpretations of the term. However, one common characteristic that is agreed on is that data lie at the core of digital twins.

To avoid the same pitfalls and resistance that we have encountered in implementing BIM in the industry, it is important to clarify what a digital twin is and the responsibilities of each party involved in the project from the outset. As with BIM, digital twinning must be viewed as a comprehensive process for managing information and data throughout the project lifespan, rather than simply a software tool.

The lack of understanding of the BIM process and the poor quality of its initial implementation have limited its trust and adoption by the industry. Additionally, the lack of client support and accountability among stakeholders for the quality of deliverables has hindered BIM adoption. These and other challenges have made it difficult for the industry to fully realise the potential of BIM and could be the same for digital twins.

It is important to recognise that digital twins might also encounter similar issues that can limit their potential. Therefore, it is essential to learn from these challenges and address them in the early stages of digital twin projects. Communication and education in the industry are critical to ensure that everyone involved understands the benefits and limitations of digital twins and how they can be effectively integrated in project workflows.

Equally important is acknowledging that, as with any new process and technology, the successful implementation and adoption of digital twins in the industry requires the support and buy-in from the client. For the project to provide real value and support the asset's operations and maintenance, it is essential that the client leads the effort and has a clear purpose in mind.

As the industry becomes increasingly interested in digital twins, we are already seeing some clients request digital twins without having any clear strategy or purpose, often confusing them with the digital model. This highlights the need for standardisation of digital twins and clear communication about what they are and what benefits they can provide. If these issues are not addressed promptly, they can lead to confusion and resistance in the industry, hindering the adoption of digital twins and limiting their potential to transform the way we manage assets.

Finally, I have observed that there is still confusion in the industry between the concepts of a digital model and a digital twin (Miskimmin, 2022). Therefore, it is important to highlight the differences between the two in order to avoid any misunderstandings.

Digital model

The digital model is essentially a static representation of the asset at a particular point in time; it does not reflect any changes that might have been made to the physical asset. The BIM process enables us to deliver a digital model to our clients, which provides a digital representation of the physical asset, along with relevant data to support its operations and maintenance. However, the digital model needs to be manually updated by the client team during the operations and maintenance stage to reflect any changes that have been made in the physical asset.

This means that the digital model is very useful in visualising the asset during the design and construction phases, but it might not be very effective for ongoing asset management.

Digital shadow

The digital shadow is a virtual representation of a physical object, created by feeding real-time data about the state of that object into a digital model.

The purpose of a digital shadow is to provide a real-time representation of the physical object, which can be used for a variety of purposes, such as monitoring, analysis or simulation.

It is essential to note that a digital shadow is not the same as a digital twin. One key difference between a digital shadow and a digital twin lies in the direction of information flow. In the case of a digital shadow, data about the physical object are fed into the model to create a virtual representation. Changes in the physical object trigger automatic updates to the digital object; however, the reverse does not occur, as updates from the digital to the physical object require manual intervention.

Digital twin

The digital twin goes beyond being just a static digital image of a physical asset, as it represents a virtual replica of the asset with a two-way connection to the physical asset. This connection enables the digital twin to receive and integrate data from the physical asset in the digital model, allowing it to be managed from the digital version.

One of the key features of a digital twin is its ability to integrate data from various sources, including sensors and cameras, in a single digital model. This model serves as a digital representation of the physical asset and enables communication between the digital and physical versions of the asset.

Furthermore, the digital twin is not limited to simple data integration and analysis; it also enables the application of advanced technologies, such as machine learning algorithms, to enhance the capabilities of the physical asset. By training machine learning models on data collected from the physical asset, the digital twin can provide insights and recommendations for optimising its performance and reducing down time. Figure 7.9 aims to support the understanding of this classification.

The format of a digital twin is not limited to just 3D models, as many people tend to think. Depending on the project's needs and objectives, various formats can be used to represent a digital twin. For example, a 2D representation may suffice in some situations, while a 3D representation may be required in others. The choice of 2D or 3D digital twin will ultimately depend on the physical asset's characteristics and the client's goals.

A 2D digital twin can be as simple as a floor plan of a building and can provide valuable insights into usage levels, energy consumption and user behaviour. This information can be used to make informed decisions about improving the layout, increasing or decreasing the number of meeting

Figure 7.9 Digital twins (Icons: Benvenuto Cellini and M.Style/Shutterstock)

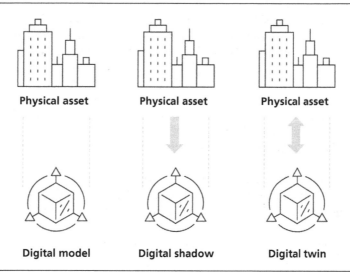

Physical asset	Physical asset	Physical asset
Digital model	Digital shadow	Digital twin

rooms and enhancing energy efficiency. By comparing the digital twin to real-world data, clients can gain a better understanding of the asset's performance and take steps to optimise it.

What to consider

Effective information management and data capture systems are essential for creating a high-performing asset during the various stages of a construction project. The industry must give appropriate attention to data capture, information management and project management, to ensure the efficient and effective implementation of digital twins. Digital assets have numerous uses, including analysing data and predicting system behaviour, but their purpose must be clearly defined from the beginning.

Similar to the Asset Information Model (AIM), to achieve the intended purpose of the digital twin, it is necessary to develop a strategy. This strategy should describe several key aspects: the data required, how often updates will occur, security measures, responsibilities, and ownership, among others. It should also outline the processes and tools for collecting, storing, and processing data.

The utilisation of digital twin technology is transforming various industries by facilitating real-time monitoring, analysis and optimisation (Forbes, 2023). Nonetheless, the efficacy of digital twins primarily depends on the quality and precision of the data used to generate them. Inaccurate or unreliable data could result in flawed predictions, erroneous analyses and, ultimately, wasted time and resources.

To guarantee the success of digital twins, it is imperative to establish a robust data gathering system that can provide precise, and up-to-date data. Moreover, it is crucial to establish a data management system that can store, process and scrutinise the gathered data. This system should be capable of handling massive data volumes, ensuring data quality and authenticity, and facilitating real-time data analysis and visualisation. With accurate data at hand, digital twin technology can be applied in countless ways.

BIBLIOGRAPHY

Abisogun B, Bailey J, Callcut M *et al.* (2021) *Digital Twin Toolkit: Developing the Business Case for Your Digital Twin.* Centre for Digital Built Britain, Cambridge, UK.

BIM4Housing (2023) BIM4Housing's vision is for all stakeholders, in the provision of housing, to maximise the value added by Better Information Management. https://bim4housing.com/ (accessed 01/12/2023).

BSI (2012) BS 8541-1:2012: Library objects for architecture, engineering and construction. Identification and classification. Code of practice. BSI, London, UK.

BSI (2013) PAS 1192-2:2013: Specification for information management for the capital/delivery phase of construction projects using building information modelling. BSI, London, UK.

BSI (2014) ISO 55000:2014: Asset management. Overview, principles and terminology. BSI, London, UK.

BSI (2019) BS EN ISO 19650-1:2018: Organization and digitization of information about buildings and civil engineering works, including building information modelling (BIM). Information management using building information modelling. Part 1: Concepts and principles. BSI, London, UK.

BSI (2020a) BS EN ISO 19650-3:2020: Organization and digitization of information about buildings and civil engineering works, including building information modelling (BIM). Information management using building information modelling. Part 3: Operational phase of the assets. BSI, London, UK.

BSI (2020b) BS EN ISO 19650-5:2020: Organization and digitization of information about buildings and civil engineering works, including building information modelling (BIM). Information management using building information modelling. Part 5: Security-minded approach to information management. BSI, London, UK.

BSI (2021) BS EN ISO 19650-2:2018 & Revised NA: Organization and digitization of information about buildings and civil engineering works, including building information modelling (BIM). Information management using building information modelling. Part 2: Delivery phase of the assets. BSI, London, UK.

BSI (2022a) BS EN ISO 19650-4:2022: Organization and digitization of information about buildings and civil engineering works, including building information modelling (BIM). Information management using building information modelling. Part 4: Information exchange. BSI, London, UK.

BSI (2022b) BS 8644-1:2022: Digital management of fire safety information. Design, construction, handover, asset management and emergency response. Code of practice. BSI, London, UK.

Cabinet Office (2011) *Government Construction Strategy.* Cabinet Office, London, UK. https://assets.publishing.service.gov.uk/media/5a78ce8eed915d07d35b2933/Government-Construction-Strategy_0.pdf (accessed 23/11/2023).

Davies H, Sharma A, Kemp A *et al.* (2021) *Building Regulations Advisory Committee: Golden Thread Report.* Ministry of Housing, Communities & Local Government, London, UK. https://www.gov.uk/government/publications/building-regulations-advisory-committee-golden-thread-report/building-regulations-advisory-committee-golden-thread-report (accessed 01/12/2023).

Department for Levelling Up, Housing and Communities (2022) Guidance: The Building Safety Act. https://www.gov.uk/guidance/the-building-safety-act (accessed 01/12/2023).

Digital Twin Consortium (2020) *Definition of a Digital Twin.* Digital Twin Consortium, Boston, MA, USA. https://www.digitaltwinconsortium.org/initiatives/the-definition-of-a-digital-twin/ (accessed 01/12/2023).

Digital Twin Consortium (2023) https://www.digitaltwinconsortium.org/ (accessed 01/12/2023).

Digital Twin Hub (2023) DT Hub: Building Better Connections. https://digitaltwinhub.co.uk/ (accessed 01/12/2023).

EC (European Community) (2016) Regulation (EU) 2016/679 of the European Parliament and of the Council of 27 April 2016 on the protection of natural persons with regard to the processing of personal data and on the free movement of such data, and repealing Directive 95/46/EC (General Data Protection Regulation). *Official Journal of the European Union* **119**.

Forbes (2023) Digital Twins: 18 Ways They Can Be Leveraged Across Industries. Forbes, 27 Nov. https://www.forbes.com/sites/forbestechcouncil/2023/11/27/digital-twins-18-ways-they-can-be-leveraged-across-industries/?sh=101cd74d58f8 (accessed 01/12/2023).

Groom R, Gleeson B, Coddington I, Bradley A and Kavanagh J (2014) *Measured Surveys of Land, Buildings and Utilities*, 3rd edn. Royal Institution of Chartered Surveyors, London, UK. https://www.rics.org/content/dam/ricsglobal/documents/standards/measured_surveys_of_land_buildings_and_utilities_3rd_edition_rics.pdf (accessed 30/11/2023).

Hackitt J (2018) *Building a Safer Future: Independent Review of Building Regulations and Fire Safety: Final Report*. The Stationery Office, London, UK. https://assets.publishing.service. gov.uk/media/5afc50c840f0b622e4844ab4/Building_a_Safer_Future_-_web.pdf (accessed 01/12/2023).

HMG (Her Majesty's Government) (1972) Defective Premises Act 1972. The Stationery Office, London, UK.

HMG (2018) Data Protection Act 2018. The Stationery Office, London, UK.

HMG (2022) Building Safety Act 2022. The Stationery Office, London, UK.

Institute of Asset Management (2014) What is AM? *YouTube*, 10 Sept. https://www.youtube.com/watch?v=el0A8js4vK8&t=21s&ab_channel=TheIAM (accessed 01/12/2023).

Miskimmin I (2022) Digital twins, shadows and models. *LinkedIn*, 29 Nov. https://www.linkedin.com/pulse/digital-twins-shadows-models-iain-miskimmin/?trackingId=%2Bm3s5PH0rpPk5mIrz9mFXA%3D%3D (accessed 01/12/2023).

Norton J (2021) Infrastructure 2021: It's time to rethink 'shovel-ready'. *Forbes*, 15 Sept. https://www.forbes.com/sites/forbesbusinesscouncil/2021/09/15/infrastructure-2021-its-time-to-rethink-shovel-ready/?sh=5ac763411681 (accessed 01/12/2023).

PwC (2020) *Assessment of the Benefits of BIM in Asset Management. Part 1: Context and Methodology*. PwC, London, UK. https://www.cdbb.cam.ac.uk/files/201117_uoc_req_2_value_of_bim_in_am_pt1_final.pdf (accessed 30/11/2023).

Stanton J (2022) BS 8644-1 Digital management of fire safety information: an information management perspective. *BIMplus*, 12 Aug. https://www.bimplus.co.uk/bs-8644-1-digital-management-of-fire-safety-information-an-information-management-perspective/ (accessed 01/12/2023).

Amador Caballero
ISBN 978-1-83549-446-2
https://doi.org/10.1680/iceedc.9446208

Chapter 8
Sharing experiences

8.1.　Introduction

In this chapter, I share my personal insights regarding the common challenges encountered when implementing BIM in projects. These insights are based on my own experiences and the challenges that I have faced over time in different projects. Addressing these challenges promptly and engaging in dialogue with relevant stakeholders are key steps in improving project delivery. This chapter offers a comprehensive discussion of these challenges, organised according to several factors: the survey process, point cloud scope and verification surveys; design model creation; information management and commercial considerations.

A recurring problem in the construction industry is the repetition of errors. This often occurs as a result of high staff turnover and poor communication between teams, which disrupts the transfer of knowledge and experience within organisations. I strongly recommend documenting both successes and challenges in BIM implementation, as discussed in Section 2.5. Learning from mistakes is essential and it's crucial to capture them to ensure that knowledge is shared and similar issues can be avoided in future projects. An important part of these lessons lies in identifying their root causes and analysing them to address complex questions. Embracing this approach enhances the potential to mitigate similar challenges in future projects.

I provide valuable insights into these challenges, aiming to offer practical advice and solutions before you come across similar problems in your own projects. While learning from your own mistakes can provide the most valuable insights, these errors can also result in various difficulties.

Before initiating any construction project, it's crucial for the team to thoroughly review challenges faced in previous projects. Taking this proactive approach will better prepare you for potential hurdles and increase the likelihood of a successful project outcome.

Each section in this chapter presents actionable tips and guidance, meticulously curated to assist you in the effective implementation of BIM. As you work through this chapter, you will not only be better prepared to avoid common pitfalls but also feel empowered to guide your team and organisation towards a successful and enriching BIM journey. You can similarly apply this approach with your own findings to support implementation within your organisation. Overall, this chapter aims to be your ally and guide, helping you navigate the often complex landscape of BIM implementation and steering you towards success.

8.2.　Surveys

The importance of completing a point cloud survey

The common issue

In construction projects, especially those involving existing buildings, it's often the case that no reliable design information is available. Sometimes, the existing information might be incomplete and, therefore, cannot be trusted for the project. Furthermore, it's common in most construction projects to need to verify the completed work on site. This is to ensure that it corresponds to the original design and to catch any errors that might cause problems later during construction. The absence of reliable data can result in a range of issues, including costly delays and the need for redesigns.

Recommended approach

An effective way to address this issue is to conduct point cloud surveys at various stages of the project. Ideally, the survey should take place after a complete strip-out, as this allows you to identify elements that are normally hidden and wouldn't be captured in an earlier point cloud survey. However, owing to time constraints in many projects, an initial point cloud survey before the strip-out is often necessary. This enables design development to proceed without waiting for the complete strip-out to be finished. Subsequently, a second point cloud survey should be carried out after the strip-out or demolition to verify the building's actual conditions, which can then inform any necessary design updates.

It's fundamental to understand that point cloud surveys shouldn't be seen as an extra cost but as an investment. When conducted properly, these surveys can help avoid additional costs and project delays, making them a cost-saving measure in the long run.

For best practice, the time and budget for these surveys should be included during the project's bidding phase and integrated in the project timeline or 'critical path'.

As highlighted, point cloud surveys are not only beneficial before and after demolition to support design development; they are also invaluable during the construction stage. These surveys are particularly useful for identifying any errors in critical work packages, such as structural elements and foundations. By validating and verifying these works and catching errors promptly, you can prevent expensive mistakes and subsequent delays, ensuring that the project stays on track. The cost of conducting these checks is minimal, compared with the consequences of not identifying errors in a timely manner, owing to the significant impact this can have on the project.

Verify client information

The common issue

Teams often assume that the information provided by the client team, completed during the early stages of the project, is both valid and accurate, and proceed to develop designs based on the unverified data. This is a risky practice, particularly for the main contractor in a design-and-build procurement project. The responsibility for data accuracy is often passed to the main contractor, making it crucial to verify all details provided.

Recommended approach

To mitigate this risk, it's essential to carry out appropriate surveys to confirm the information supplied by the client. In cases where the client is willing to assume responsibility for any issues arising from an inaccurate design, or in a traditional contract procurement, the need for verification may be less stringent. However, this is rarely the case, and the responsibility usually rests with the main contractor.

For projects in which you, as the main contractor, are responsible for the design, you must rigorously validate and verify all client-provided information, including 2D and 3D designs. This is crucial, especially for refurbishment projects, to ensure that the design accurately matches the existing conditions of the building. It is essential to compare the 2D or 3D designs against the point cloud data. Identifying discrepancies early on can prevent delays and additional costs, making this a crucial step in the design process.

Don't be lulled into a false sense of security if you're told the design is based on point cloud data. Errors can occur during the design process, leading to plans that don't accurately represent the project's needs.

Your role involves accountability for any issues that may arise during construction. Therefore, you should not assume responsibility for a design without thorough verification, as failure to do so could result in costly errors during construction.

Additionally, it is important to ensure that the 2D and 3D information match correctly. For example, I came across schematic information that contradicted the information available in the 3D design. All project information should be coherent.

Clearly outline the scope of the point cloud survey

The common issue
Individuals or organisations new to the world of point cloud surveys often encounter a pervasive issue: they often approach survey companies for quotations without a well-defined scope of work or without clarifying the required outcomes of the point cloud survey. This lack of preparation can lead to wildly different cost estimates and, ultimately, deliverables that do not meet the project's needs. Additionally, poor communication and vague expectations can further muddy the waters, leading to project delays and cost overruns.

Recommended approach
- *Thoroughly define the scope beforehand.* Before seeking quotes, it's crucial to draft a comprehensive scope of work. This should cover not only the areas to be surveyed but also the expected level of accuracy, the survey methodology and the specific deliverables required, such as drawings, models and other forms of documentation and their formats. Doing this increases the chances of receiving accurate and comparable quotes from different firms.
- *Conduct a collaborative delivery team meeting.* Organise a meeting with your design and delivery teams and survey consultants. Use this opportunity to openly discuss everyone's needs, expectations and potential concerns when defining the scope. This proactive approach minimises the risk of assumptions sneaking into the design and allows for more accurate cost estimation.
- *Be selective in modelling elements.* In scan-to-BIM projects, not every element needs the same level of geometric detail. You can save time and money by carefully choosing which elements require a detailed representation of existing conditions and which can be depicted more generically while still being accurate. This is a best use of resources and enhances the overall design and coordination of the project.
- *Network and partner with trusted consultants.* Cultivating a network of reliable consultants can yield long-term benefits. Not only does this enable you to secure trusted services quickly, but it also facilitates early engagement in future projects, helping you sidestep common setbacks and delays.
- *Learn and adapt from previous engagements.* Your past experiences with survey companies can offer valuable insights. Reflect on how well previous projects went. Use this information to guide your choice of consultants for upcoming projects. Additionally, consider the lessons learned from past surveys to avoid repeating mistakes. Questions to ask might include: Was the survey accurate enough? Was it delivered on time? Were all areas accessible to complete the survey? Did the model reflect the correct graphical details needed for design coordination? Were you satisfied with the verification report? Were there any areas for improvement?

- *Consider structured agreements.* In lieu of commercial retainers, negotiate a structured agreement with your survey consultant to mitigate potential delays or substandard deliverables. This meets with the CLC and NEC guidance (NEC, 2022) on moving away from retentions, encouraging a framework that promotes timely, quality work while providing a clear resolution pathway if issues arise.

Verify the survey deliverables

The common issue

A point cloud survey can yield invaluable information about a building. However, a common issue is the tendency to assume that the output from the survey consultant is both accurate and comprehensive. Often, the delivery team neglect the responsibility to verify this information on site, and may also fail to check that it corresponds to all the elements outlined in the project scope. This oversight can result in errors and missing details. Addressing these issues can be not only expensive but also time-consuming, impacting both the budget and the project timeline.

Recommended approach

- *Consistent and regular check-ins.* Ongoing communication is necessary for identifying potential issues during the survey and clarifying any ambiguities with the survey consultant. Establish a schedule for regular check-ins with the survey consultant throughout the survey's execution. These could range from weekly updates to more frequent touchpoints, depending on the complexity and pace of the project.

- *Deliverables review meeting.* It is crucial to review the final work to confirm that it fulfils the delivery team's needs. I recommend meeting the point cloud survey consultant and the delivery team to review the survey and deliverables, addressing all questions or concerns openly. This allows for early identification of any issues or additional information needs, thereby avoiding future delays.

- *Mandatory on-site verification.* The lead appointed party shall conduct on-site checks to verify the survey data. This step serves as a safety net, catching any inaccuracies before they escalate into costly errors. Given the importance of survey information for smooth design and construction progress, any effort to minimise risk and improve the quality of the data will pay dividends during project execution.

- *Before finalisation.* Upload the deliverables to the common data environment and communicate this to all current and future delivery team members, so they can access and review the information as required. If you're using a model viewer separate from the common data environment, make sure that the current team have access to review the model. The team needs to provide the necessary comments for its amendment before the work of the survey consultant is concluded.

Avoid duplicated survey information

The common issue

In construction projects, it's vital for the delivery team to use a consistent set of data, which could include such elements as control networks, grid lines and floor finish levels. However, lapses in information management and communication frequently lead to duplicate surveys and inconsistent information. If separate surveys are conducted and inconsistencies arise, these should be immediately flagged with the site lead for resolution. These inconsistencies not only cause confusion but also result in delays, additional work and extra costs, jeopardising the overall success of the project.

Recommended approach

To avoid these pitfalls, it's essential to establish clear communication channels and robust information management protocols across the delivery team. Task teams that conduct their own additional surveys should be aware if any prior survey information is available, and the existence of a control network.

Furthermore, it's advisable for task teams to regularly review the levels and data being used, especially during the early stages of the project. By taking this proactive approach, potential problems can be identified before they become major issues that lead to delays and additional costs.

Model responsibilities for retained services and structure

The common issue

To effectively coordinate new elements in refurbishment projects, accurate modelling of existing structures and services is crucial. Difficulties often arise when retained services are incorporated in the design or when previously unseen structures are discovered as the project progresses. Typically, task teams aren't initially responsible for updating designs to include these existing elements, leading to delays, miscommunication and finger-pointing between teams. Frequent design changes further complicate the situation, requiring continuous updates to the survey model.

The current approach has significant drawbacks. It usually relies on point cloud surveys and involves delegating the modelling of retained services to survey consultants, who follow directives from M&E consultants and structural engineers. This process becomes problematic when decisions about retaining certain elements change during the design phase. Constant updates to the survey model are then needed, making the task laborious and time-consuming. This can also serve as an excuse for task teams to justify delays, perpetuating a culture of blame concerning the survey model's quality and accuracy.

Recommended approach

To manage this complex challenge more efficiently, I strongly recommend that the task of modelling any retained services and structural elements be delegated directly to the task teams, spanning both the design and construction phases of the project. This proactive strategy offers several benefits. First, it ensures that newly discovered elements exposed during the construction phase are swiftly and accurately incorporated in the design models by the relevant task team. Second, it decreases the likelihood of disputes arising over the accuracy of the survey model, thereby expediting the process and preventing unnecessary delays.

Incorporating this strategy in the scope of work for the task teams – and communicating it effectively – offers another advantage: it enables greater flexibility for design changes. Revisions or newly discovered elements can be quickly integrated in the model without waiting for the survey consultant to update the survey model.

When formulating the scope of works for refurbishment projects, it's essential to consider this approach. While the lead appointed party may still conduct the initial point cloud surveys and scan-to-BIM process, the responsibility for updates should be included as part of the task team appointments to mitigate the risk of delays and misunderstandings or a culture of blame. The capabilities of the task team will play an important role on this strategy.

8.3.   Design models

Fix coordinates and IFC issues from the outset

The common issue

There is still a key problem when teams set up their design models, as teams often overlook the crucial step of setting correct coordinates. This oversight can lead to a misalignment in the design between the various disciplines involved, causing problems in overall project coordination. If team members work in isolation without cross-verifying coordinates, this can result in delays and management challenges during the coordination phase. The issue is further complicated when different authoring tools are in use and when the project depends on IFC for design alignment. If an IFC model isn't positioned accurately during the import process, time-consuming manual adjustments are needed, making it crucial to address this issue promptly.

Recommended approach

To alleviate these problems, it's essential to confirm and set the project coordinates at the beginning. The team should reach a consensus on a standardised approach for setting up these coordinates, which should generally originate from survey data or, if survey data are unavailable, from the architectural design.

Teams should obtain the coordinates from the agreed discipline and make sure that these coordinates are properly configured in their authoring tools before moving forward with detailed design tasks. It's also crucial to test the coordinates with the agreed federated model viewer to ensure that they are consistent across all disciplines involved. Taking these proactive steps can help avoid costly and time-consuming errors later. By making the effort to validate coordinates early on, you can make the design process more efficient, avoid unnecessary delays and cut project costs, ultimately leading to a more successful project outcome for all involved.

Effective information exchange to support the design

The common issue

In many projects, the issue we often encounter is that task teams fail to provide the necessary information both on time and at the quality level we expect. This causes delays in coordination and design development, affecting the overall project schedule and the quality of the work completed.

Therefore, it's crucial for teams to be accountable for delivering information that's not only on time but also of the right quality. This means making the necessary adjustments and addressing any comments promptly, to keep everything on track.

Recommended approach

To tackle this issue, both the design manager and the technical services manager play crucial roles. They can take several actions to solve the problem.

- *Set clear deadlines.* Both managers should agree on specific deadlines for sharing information among team members. Sticking to these deadlines ensures that the project stays on track.
- *Ensure quality and timeliness.* It's not enough just to get information on time; the task teams also need to ensure that the information is accurate and complete. Any changes or comments from previous versions must be quickly incorporated to maintain high-quality work.

- *Be proactive, not just reactive.* The managers should identify potential problems and bottlenecks that could arise during the project. They should take preventive steps and prioritise resolving comments based on the project schedule and the needs of other teams.

- *Enforce accountability.* If teams miss deadlines or provide poor-quality information, the managers need to take action. This could mean revisiting the project contract or even imposing penalties on those who consistently fail to meet expectations.

- *Keep communication open.* Regular meetings should be scheduled to discuss the status of information sharing. This maintains transparency and ensures that everyone is working towards the project's goals.

By taking these steps, the design manager and technical services manager can significantly improve coordination. This will reduce delays and make the development process more efficient, contributing to the project's overall success.

Consider the impact of slab deflection

The common issue
In existing buildings, it's crucial to consider the impact of slab deflection during the design phase of a project, especially when ceiling height restrictions are in place. Failure to account for these deflections can result in significant issues during construction, such as difficulties in coordinating services within the ceiling void or the need to lower the ceiling level, going against both the initial design and client requirements.

Recommended approach
To circumvent such issues, it's essential to include an assessment of slab deflections as part of the design coordination process. The survey consultant should produce a heat map to illustrate these deflections. The data should then be integrated in the design model by the structural engineer and coordinated with both the architect and the M&E consultant. This step is vital for ensuring that the design is accurate and corresponds to the building's actual structural condition.

By incorporating the heat map findings in the design model, you can prevent potential issues during construction. Taking slab deflections into account will result in a design that accurately reflects real-world conditions, thereby facilitating effective coordination between all parties involved in the construction process. It's important to specify this requirement when appointing the survey consultant and structural engineer to guarantee successful design coordination.

Coordinate hangers above the ceiling

The common issue
When you're working on projects that have tight ceiling voids filled with numerous services (pipes, ducts, detectors, etc.), it's easy to underestimate the importance of properly placing hangers. These hangers not only support the ceiling but also help in organising the services running above it. In environments where the ceiling height is already restricted, neglecting hanger placement could lead to poorly coordinated design and complications in construction.

Recommended approach
To avoid these issues, it's important that task teams pay close attention to the hanger locations in their design models. This level of graphical information should not be overlooked, as it can significantly affect the coordination of services above the ceiling. Hangers are an element that not all task

teams consider important or choose to include within their design models. Hence, it is important to ensure that they are included at the correct stage of the project.

This scenario should be considered as part of your risk register when planning the design with the task teams.

Only accept drawings from the coordinated model

The common issue

A common issue arises when task teams lacking adequate BIM skills resort to traditional CAD methods to produce 2D drawings. These drawings are then passed on to a BIM specialist, who creates the 3D model. This method is inefficient and leads to duplicated efforts and information, which in turn slows down the flow of essential project information. Most concerning is that the 2D information uploaded to the common data environment (CDE) is not extracted from the coordinated 3D model; instead, the original 2D drawings provided to the BIM consultant are uploaded. This results in a lack of design coordination and inconsistencies between the 2D drawings and the coordinated 3D model.

Recommended approach

To address these issues, it is fundamental that all project drawings be extracted directly from the coordinated 3D model. Only drawings produced in this manner should be uploaded to the CDE and subject to the approval workflow. Any other drawings should not be uploaded to the CDE, or they should be rejected outright.

To ensure this requirement is met, clear communication with consultants and subcontractors is essential during the tender stage, and this should be part of the BIM Execution Plan. This will ensure that everyone involved understands the necessity of avoiding contradictory information and the importance of using only drawings extracted from the coordinated 3D model. Additionally, it is recommended to check that the drawings are being created within the BIM authoring tool. This gives the team extra confidence that the correct approach is being followed and ensures that the 2D and 3D information match correctly. This approach can enable a more efficient BIM process and guarantee that accurate and consistent project information is used throughout the construction phase.

Reject construction drawings with unresolved model issues

The common issue

Design drawings must be exported from 3D models to create the 2D versions that are uploaded to the common data environment for the appropriate approval workflow. However, when teams rely solely on the 2D information exported from the 3D model, they run the risk of overlooking issues. This is because relying solely on 2D information for design coordination has its limitations. Issues might not be evident in the 2D drawings but could still be active and visible within the 3D environment, leading to an uncoordinated design and complications during the construction stage.

Recommended approach

- *Multidimensional review.* All members of the delivery team should engage not only with the 2D drawings but also with the 3D model to ensure comprehensive design coordination. This multidimensional approach enables the team to identify and address issues that might not be apparent in 2D formats.
- *Frequent checks.* Different task teams, led by the lead designer, should engage in regular dialogue and perform the necessary 3D model reviews, starting from the early design

stages. They should also clearly communicate the status of the design. This proactive approach ensures that the design is fully coordinated, thus minimising risks during the construction phase.

■ *Selective resolution.* I'm not implying that all issues in the model need to be resolved simultaneously. The focus should be on resolving issues specific to the 2D drawings for particular zones that are uploaded for approval within the CDE.

■ *Stakeholder engagement.* A key challenge lies in ensuring that all task teams update their design models and drawings to accommodate any corrections or modifications. Effective communication and accountability measures can help overcome this challenge.

By implementing these measures, it is possible to significantly reduce the risk of encountering problems on site caused by inaccuracies in 2D drawings originating from an uncoordinated 3D model.

8.4. Information management

Establish the correct workflow of information approval

The common issue

For many organisations, a significant challenge arises when using the common data environment (CDE). Often, the approval process for information, as well as team responsibilities, is unclear. This causes teams to misuse the system, resulting in information being held up, which leads to delays in coordination between team members. Such delays can disrupt project timelines and impair overall performance.

Recommended approach

To address this issue, it's crucial to establish a clear and efficient workflow for approving information within your organisation's CDE. Your team needs to receive appropriate training and take responsibility for ensuring a smooth flow of information within the workflow. This will ensure that all offices and teams follow the same procedures. While some minor tweaks may be necessary based on client input, which should be agreed at the beginning of the project, the overall approach should be consistent across all projects. The information management team should therefore be responsible for setting up this workflow, taking into account the needs of both the design manager and the client.

The following steps can help to make the workflow more efficient.

■ *Customise to business needs.* Align the workflow with specific business processes and project requirements.

■ *Time allocation.* Ensure that the workflow allows enough time for the project team to review the information and offer comments, as laid out in contractual agreements.

■ *Information filtering.* To avoid overwhelming people and causing unnecessary delays, distribute only relevant information to those responsible, for review and comment.

Manage information on the CDE correctly

The common issue

While setting up the CDE with the appropriate workflows and the right naming conventions is a step in the right direction, it doesn't guarantee a smooth flow of information in the project. Proper training for all team members is essential, and everyone must be held accountable for their tasks to ensure successful project delivery and avoid delays in information delivery.

Recommended approach

To ensure a well-executed project, consider implementing the following measures.

- *Adequate training.* Ensure that all project team members are adequately trained, as soon as they join the project, to manage and input information correctly in the CDE.

- *Accountability.* Hold everyone accountable for their respective tasks and ensure that deadlines and quality standards are met. Clearly establish expectations for each member of the project team.

- *Clear quality expectations.* Establish and enforce quality standards for information to prevent the submission of poor-quality work just to meet deadlines.

- *Timely feedback.* The project team must provide their comments and take the necessary actions as part of the workflow approvals within the agreed time frames to avoid delays.

- *Regular reports.* Utilise information management reports to assess stakeholder compliance with procedures and identify those requiring guidance. Enhance transparency about information status within the CDE to proactively address performance issues and keep the project on track.

- *Task information delivery plan (TIDP).* Hold teams accountable for the TIDP, established at the time of the appointment, to ensure timely and accurate information delivery. If updates are necessary during the project, these must be communicated promptly to the lead appointed party.

Complying with information exchange methods

The common issue

Occasionally, especially during the execution of works, task teams bypass the agreed workflow for the approval of information. They do this by sharing information by email, under the assumption that this is a faster way for the information to reach the relevant parties. However, this approach introduces risks and leads to poor handover of information, as well as contractual and regulatory issues.

Recommended approach

It's vital to adhere to the proper information exchange protocols as outlined in the appointments. Any deviations from the agreed method of exchanging information should not be tolerated by any member of the project team. This is because of the implications it will have, not only during the project execution but also afterwards. To avoid such scenarios, it's important that the entire project team is committed to using the CDE within the appropriate time frames. This will ensure a smooth and correct flow of information and avoid potential penalties.

Use only accepted construction information on site

The common issue

In many construction projects, I've noticed that teams occasionally use preliminary information from the CDE that hasn't been officially issued for construction. This often happens because task teams are satisfied with the content and, therefore, avoid going through the process of reissuing the information. However, this practice can result in costly mistakes and lead to contractual issues for the business.

Recommended approach

To prevent such issues, it's crucial for all team members to use only those drawings designated for construction, which have gained the published state, through the CDE workflow. If you find yourselves waiting for a drawing to be revised, reviewed and accepted, it's wiser to wait than to forge ahead with a preliminary version.

Even if certain task teams provide preliminary information and no objections are raised by the rest of the project teams, it remains your responsibility to ensure that the information attains the appropriate contractual status by being authorised and accepted. No preliminary information should be used on site.

Furthermore, it's crucial to emphasise that supplying preliminary drawings to the appointing party during the handover phase is not acceptable. This practice falls short of the requirement to provide only the most recent, authorised and accepted construction issue drawings.

Stick to the accepted design

The common issue
A common issue in construction projects is that subcontractors often deviate from the authorised and accepted design when executing work on site. This deviation may be intentional, to make the construction process easier in their view, or unintentional, either through the use of outdated design revisions or by misinterpreting information. Such deviations can have serious implications, not only disrupting the project schedule but also causing conflicts between the different trades involved in the project. Furthermore, failure to adhere to accepted design specifications can result in legal issues and a less-than-satisfactory handover to the client.

Recommended approach
To address this issue, it's crucial for the lead appointed party to rigorously monitor and manage the construction process, and to ensure that all stakeholders strictly adhere to the authorised and accepted design information and follow the correct approval procedures.

To mitigate these risks, the lead appointed party should:

- Ensure that all stakeholders use up-to-date, authorised and accepted design models and drawings before starting work on site.
- Closely monitor the work, to identify any deviations from the authorised and accepted design and address them promptly.
- Confirm that the project model and drawings accurately reflect any design changes and ensure that this information is communicated to all stakeholders.

Use digital tools for model and design review

The common issue
Despite the advances in digital construction technology, some individuals within the delivery team are still hesitant to use digital tools to review models and design elements in a 3D environment. This reluctance means they miss out on critical opportunities for better understanding the design and identifying potential issues before they become costly errors during construction.

Recommended approach
To address this issue, it's essential for all delivery team members to receive training and support in using digital tools as soon as they join the project. These digital solutions allow for the review of drawings and models from any location, eliminating the need for large laptops to access the model. This enables a clearer understanding of the design and allows for comparisons of the design against the works to be made while walking around the construction site.

Digital tools also promote better coordination and collaboration between all team members. This aids not only in the early identification of errors but also in their timely resolution, facilitating the creation of high-quality coordinated drawings essential for the construction phase.

By implementing digital tools for design review and task management, the delivery team can ensure smoother and more efficient project delivery. This minimises the risk of costly mistakes and delays. Therefore, it's crucial for all involved parties to prioritise support for the adoption of digital tools, paving the way for a more streamlined and effective project lifecycle.

Manage task teams to deliver asset data

The common issue
One of the key challenges in information management is the timely and accurate delivery of asset data, often structured in COBie format. Teams frequently leave this crucial task until the project's end, leading to delays and potential inaccuracies. Additionally, not all teams are experienced in delivering this type of data, causing further complications.

Recommended approach
- *Early planning.* If it is a project requirement to provide asset data, begin gathering this information during the design and construction phases, rather than waiting until the end of the project. The BIM Execution Plan must clearly specify these requirements and the lead appointed party should not permit teams to miss data delivery at the appropriate stages.
- *Periodic reviews.* Carry out regular checks throughout the project to identify any errors at an early stage and ensure that the data delivery is accurate and on time.
- *A multiplicity of task teams.* Various task teams are involved in data delivery, with the mechanical and electrical teams often having the most extensive obligations. It's essential to define clear roles and responsibilities for data delivery from the outset.
- *Data integration.* Some task teams are solely responsible for issuing certain types of data. To avoid misunderstandings later on, it's crucial to establish the process for integrating these data in the design model during the tender stage. During the tender stage, it's crucial to clarify the complete management of the data that will be provided by various task teams.
- *Commercial support.* Fully involve the commercial team to ensure smooth communication with the task teams. Implement any necessary commercial actions to guarantee timely delivery of the information.

Validate and verify the asset data

The common issue
While task teams are each responsible and accountable for delivering accurate asset data as part of their roles, the lead appointed party cannot take it for granted that the information provided by these teams is correct, as inaccuracies in the information supplied by task teams are not uncommon.

Recommended approach
It's essential for the lead appointed party to validate and verify the information from each task team at every stage of the project, ensuring that it fulfils the data requirements outlined in the BIM Execution Plan. This guarantees the accuracy and completeness of data before acceptance. The digital team has the skills to validate asset data in COBie or any other structured format chosen by the client, ensuring that the data are correctly formatted and meet the required specifications. However,

they are not the appropriate team to verify the content's accuracy and its correspondence with the design and work completed on site. Therefore, it's the technical team responsibility to verify all information at each stage, including manufacturer information, supplier details and warranties.

A collaborative multidisciplinary approach is essential for meeting data format standards and ensuring the accuracy of information at various project stages.

Improve the delivery of O&M documentation

The common issue

In the construction industry, there's often a shortfall in delivering accurate operation and maintenance (O&M) documentation to clients. Typically, this task is pushed to the end of the project, when teams are under pressure to complete the work. Consequently, this crucial task doesn't get the attention it deserves. This outdated approach needs revising, not only to comply with new regulations but also to better support clients and safeguard the business.

Recommended approach

Discussing and agreeing on the O&M content with the client early in the project will help ensure a smooth transition from the construction phase to the O&M stages.

Clear roles and responsibilities should be defined within the lead appointed party. This clarity allows that all information issued by each task team is reviewed, validated and verified, based on the design and work completed on site.

Education is key; it's crucial to emphasise the importance of providing accurate information to the client. Failure to do so can have significant implications for both individuals and the business.

To ensure the effective delivery of accurate O&M information, it's recommended that this task is integrated throughout the construction project, rather than postponed until the end. This approach affords the lead appointed party sufficient time to review all relevant documents, validate their format and verify their accuracy.

In my opinion, all project-related data should be shared through a common data environment, using a well-defined approval workflow to ensure traceability. Even if some firms choose to outsource the collection of O&M information, the lead appointed party should still approve the data within the project's CDE. This step guarantees traceability and accountability before the data are handed over to a third party, if necessary.

This proactive approach reduces contractual risks, fosters efficient collaboration and contributes to a successful project outcome for all parties involved.

8.5.   Commercial considerations

Manage client's expectations at tender stage

The common issue

A recurring issue in project management is the lack of clarity in client expectations and requirements during the tender stage. This can lead to misunderstandings, additional costs and complexities as the project progresses.

Recommended approach

To avoid these issues, it's necessary to have open and honest conversations with the client at the tender stage. This allows any assumptions to be addressed and enables the lead appointed party to produce accurate documentation. It also helps in communicating the correct requirements to the potential task teams, ensuring the best possible tender submission.

Here are some areas to consider.

- *Clarify roles and responsibilities.* Be sure to clarify the roles and responsibilities of the appointing party. Discuss how the appointing party will be involved in the project, the correct use of the common data environment (CDE) and the participation in workflow approval. Understand their preferences for operations and maintenance (O&M) delivery and the content of the asset data. Address any concerns that might lead the task team to raise fees or have doubts about project delivery.

- *Address unrealistic expectations.* Clearly discuss expectations and requirements with the appointing party. Let them know if some of their expectations could potentially incur extra costs or are impractical based on the current project status. For example, if a point cloud survey is requested to verify construction works, discuss its timing and necessity, as such a survey might require extensive coordination and could affect the overall schedule.

- *Efficiency.* Evaluate whether the client's approach is practical, or if there are more efficient ways to achieve the desired outcomes.

- *Data collection.* Be explicit about the data needed, particularly for facility management, and address the security requirements for managing sensitive information. These requirements are sometimes mentioned in the documentation but are often not well-defined.

- *Asset tagging.* Gain clarity on asset tagging requirements. Discuss which elements require tags, who will provide them and where they should be placed. This is another requirement that is often vaguely defined in documentation, and clients frequently deviate from their original expectations.

Review the BEP and clarify BIM expectations at the tender stage

The common issue

A prevalent issue is that the different task teams frequently fail to adequately review and understand the BIM Execution Plan (BEP) or communicate their input for improving it. On many occasions, the document either goes unreviewed or encounters resistance in terms of compliance. Task teams often fail to provide the constructive feedback and collaborative approach that are essential for the effective implementation of BIM.

Recommended approach

To guarantee that the task team understands the BIM Execution Plan (BEP) correctly and collaborates to enhance it, the following steps should be conducted.

- *Meticulous review.* During the tender stage, task teams should rigorously review the BEP, as well as any related documentation, before providing tender submissions. Make sure that the information provided is acknowledged by the recipient.

- *Clear understanding.* All stakeholders should strive for a thorough understanding of the BEP. To facilitate this, consider organising a workshop to review the main content and address any questions. Transparent communication is crucial, especially for clarifying BIM-related expectations with task teams. This includes not just the documentation but also other available information, such as drawings, specifications and models. Poor communication can result in unforeseen costs or even jeopardise BIM implementation in the project. To prevent this, address any concerns promptly to ensure that all parties understand the project's objectives. Encourage teams to provide constructive feedback to effectively refine the plan.

- *Detailed responsibility matrix.* The BEP should feature a well-defined responsibility matrix that outlines the roles and responsibilities of each stakeholder. This matrix should account for each party's scope of work and clearly specify what deliverables are expected.

- *Contractual obligations.* It's important to incorporate the BEP and any accompanying BIM documentation into appointments. This ensures that everyone is contractually bound to meet the BIM requirements of the project.

Assess BIM capabilities at the tender stage

The common issue
A common pitfall in the construction industry is that supply chain selection is often driven by cost rather than capability. The result is that you might end up with task teams who lacks the skills needed to meet the BIM requirements of a project. The gravity of this issue cannot be overstated, as choosing the wrong supply chain can lead to additional costs and a host of other complications that could have been avoided.

Recommended approach

- *Early BIM assessment.* At the tender stage, every party involved in the project should complete a BIM capabilities assessment. This ensures that you're collaborating with companies that possess the essential skills for the project's BIM objectives. I recommend conducting interviews and having conversations to understand each team's culture and approach to digital construction, as assessments can offer a limited picture of each team's capabilities.

- *Past performance review.* Consider the past performance of each task team based on previously completed projects. This gives you a more comprehensive understanding of their expertise.

- *Careful selection.* As mentioned, do not choose your task teams based solely on cost, but also by considering their capabilities, culture and proven performance. This enhances the likelihood of seamless design coordination, data delivery and information management. As a result, the project is more likely to stay on schedule and within budget. Any task team attempting to charge additional fees for BIM implementation should be critically reviewed, and the reasons for the extra cost should be understood.

- *Regular communication.* Provide consistent updates to all stakeholders to ensure consistency and the timely identification and resolution of any issues. This practice minimises the risk of delays and unforeseen expenses and fosters a collaborative approach.

- *Capture team performance.* Create a database, recording the performance of each task team. This database will be invaluable when selecting task teams for future projects.

BIM requirements are non-negotiable in appointments

The common issue
I have occasionally found that, even when BIM requirements are clearly outlined, task teams sometimes skirt these essential obligations from their appointment. This can happen in several ways, such as not delivering COBie data, ignoring agreed naming conventions or avoiding design coordination duties. Once the project has started, correcting these omissions can be both difficult

and costly. Failure to meet BIM standards might require the lead appointed party to bring in extra resources. In the worst-case scenario, it might be necessary to hire BIM consultants to fill the gaps, undermining the advantages of a well-implemented BIM process.

Recommended approach

To tackle this issue, careful planning is needed during the tender phase of the project. Here are some suggestions to consider.

- Before finalising any appointments, make sure that the task teams fully understand and agree with the BIM requirements of the project.

- During the appointment process, carry out a detailed review of all responsibilities to ensure they meet the overall goals of the project and that no responsibilities have been discarded from the scope of works.

- Be cautious in choosing your project partners. A company who is hesitant to take on BIM duties might not be the best fit for your project. Opting for a less expensive contractor who lacks the necessary skills can negatively affect the project in terms of timeline, budget and quality.

Complete TIDP on time for appointment

The common issue

It's frequently observed that task teams are hesitant to create a task information delivery plan (TIDP) during the tender stage. However, this document is an essential component for effective project management. Too often, TIDPs are misused, being treated merely as drawing registers, with files uploaded to the common data environment (CDE) just before issuance. This approach neglects the intended function of the TIDP, leading to such issues as delays, miscommunication and discrepancies between project expectations and deliverables.

Recommended approach

To address these challenges, it's crucial to include the TIDP as part of the tender process. Task teams should each outline their expected deliverables in the TIDP. This should then be agreed with the lead appointed party before it's included in the appointment. Providing such clarity from the start paves the way for smoother operations and minimises the risk of delays or misunderstandings down the line.

While the TIDP may require updates as the project progresses, having a solid plan in place from the outset makes it easier to implement any subsequent changes. Companies may be reluctant to adopt this approach initially, perhaps owing to unfamiliarity or perceived inconvenience. However, the long-term advantages make it worthwhile.

One aspect that needs further clarification within the TIDP is the 'delivery date'. It's important to clarify with your task teams when the first issue is expected and when the final approval of information should occur. Clear delineation of what is expected by each delivery date prevents teams from submitting incomplete or substandard information simply to meet deadlines, thus ensuring that the quality of the project is not compromised.

Working with external BIM consultants

The common issue

When task teams who lack the necessary BIM capabilities are brought into a project as a result of contract novation, or for other reasons, and have to rely on external BIM consultants to meet specific requirements, this often leads to delays in the design schedule. Such partnerships can result in communication issues, information delays and mistakes that hinder the project's timeline and quality.

Recommended approach

A task team who require the support of a BIM consultant should conduct a comprehensive assessment of the capabilities of potential BIM consultants before beginning the collaboration. This assessment should be based on the project's needs, as well as the consultant's expertise and capacity to deliver the work on time.

Once a suitable BIM consultant has been selected, a clear and well-defined collaboration strategy should be put in place between the appointed task team and the BIM consultant. This should include robust communication protocols, regular progress reviews and close performance monitoring of the external team. Both internal and external teams should work cohesively, as if they were one unified company. Adopting this approach will minimise unnecessary duplication of work and help to keep the project on track.

BIBLIOGRAPHY

NEC (2022). *NEC and CLC Guidance for Dealing with Retention Payments Under NEC3 and NEC4 Contracts*. NEC, London, UK. https://www.constructionleadershipcouncil.co.uk/wp-content/uploads/2022/11/NEC-and-CLC-Guidance-for-Dealing-with-Retention-Payments-Under-NEC3-and-NEC4-Contracts-15.11.22.pdf (accessed 28/11/2023).

Amador Caballero
ISBN 978-1-83549-446-2
https://doi.org/10.1680/iceedc.9446209

Chapter 9
Frequently asked questions

This chapter is a supplementary guide, providing brief answers to frequently asked questions that I have encountered. Mirroring the structure of previous chapters, it aims to elaborate on key topics that were previously discussed, by addressing questions related to each area. The chapter's purpose is not only to offer quick and useful information but also to enhance the understanding gained from earlier sections.

9.1. Chapter 1: Introduction

What are the key considerations for digital transformation?

Conducting a successful digital transformation requires a strategy that includes clear governance and an accountable decision-making structure. This transformation isn't merely a task for the IT or specialised digital department; it's a cultural shift that requires the engagement and buy-in of the entire organisation. From your peers to your directors, everyone must take collective responsibility for this transformation. To achieve this, you must have the respect of these individuals and be viewed as a trusted adviser.

People are at the heart of digital transformation. Such a transformation involves winning hearts and minds, resolving conflicts constructively and communicating the transformation's purpose and benefits. Emotional intelligence is required, as it helps in understanding and addressing people's concerns, and in engaging individuals by considering their personal motivations.

Effective communication is key to aligning an organisation's overall purpose and strategy with its digital transformation approach. It's essential not only to clarify 'what' needs to be done but also to explain 'why'. Providing a compelling reason for the need for change and how individuals can contribute to the organisation reinforces their understanding that they are working towards the organisation's vision, rather than just working within it. The goal is to shift away from a feeling of being forced into a shared sense of purpose, transforming potential obstacles into enthusiastic supporters.

Additionally, motivation and incentives should not be overlooked. People naturally resist change if they don't see a benefit for themselves, such as simplified day-to-day tasks or opportunities for career growth. However, by aligning the right incentives with organisational goals, staff will be better motivated to adopt new processes and technologies. This makes it easier for people to embrace new ways of working and creates an environment that encourages learning and innovation. Therefore, for any new initiative, like BIM, to be truly effective, it's crucial to foster a culture that prioritises people and shared objectives.

What has been the most challenging obstacle you've encountered on your journey to implementing BIM?

The most significant challenge I've faced in implementing digital construction is resistance from both internal and external team members. As we aim for greater efficiency and improvements, changes in our internal processes, in the criteria for selecting external teams and in the roles and responsibilities of our internal teams become inevitable. Unfortunately, these changes aren't necessarily welcomed across the business.

While we can manage resistance from external teams by carefully choosing and collaborating with companies that share our commitment to digital construction, dealing with internal resistance is more complicated. This is often due to an unwillingness to change, despite the potential benefits.

The importance of nurturing the right culture within the business cannot be overstated. This culture should promote a genuine willingness to learn, an openness to change and an ongoing commitment to improve processes. It's essential that this desire for change is authentic and that individuals are held accountable for their actions. Support from directors and senior leaders is key to ensuring that team members not only undertake training but also fully engage with new processes and digital tools. They must be adaptable in their roles and committed to meeting their new responsibilities.

To truly realise the benefits of digital construction, an open-minded approach and a willingness to embrace change are indispensable. Without the board's commitment to this strategy, and the dedication of internal teams, the implementation of digital construction remains a consistent challenge and is likely to fail.

Can current technology support and enhance digital accessibility for all users?

In the past, access to digital models was largely limited to high-powered laptops and complex tools that were difficult for non-specialists to navigate. This posed significant barriers to entry; only the most robust computers could handle the required processing power, and users often needed extensive training. However, the landscape has changed dramatically, especially in the construction industry. Ongoing technological advances, coupled with the emergence of innovative companies and improved design practices, have made these models more accessible.

Considering model viewers, many previous issues have been resolved. Cutting-edge technology now enables us to review design models on site using mobile devices. This not only saves time but also increases efficiency. Such progress is crucial, as it grants teams easier access to models and offers greater flexibility in terms of where and how they work. This newfound flexibility can foster better collaboration and ultimately lead to improved project outcomes.

This shift goes beyond just improving access to models. Importantly, the new wave of technologies entering the industry has been designed with user-friendliness in mind, understanding that, for widespread adoption, ease of use is key. These technologies do not require specialist knowledge or the latest hardware to support project delivery. While there are certainly some new tools that demand specialist training, these are tailored for specific individuals and are not intended for mass implementation within project teams. This balanced approach ensures that the broader team can benefit from technological advances, while specialists can dive deeper, with tools designed specifically for their needs.

What is the difference between digital construction and BIM?

As discussed in Section 1.2, BIM is an enabler of digital construction, which has a broader scope. Effective information management and collaborative design in a data-rich 3D environment enable the integration of various other technologies and processes. These technologies and processes, in turn, benefit from the groundwork laid during BIM implementation.

9.2. Chapter 2: Business transformation

What are your tips for starting BIM implementation?

These are my personal thoughts and considerations on implementing BIM within a business context. It's important to note that while these suggestions can provide general guidance, individual businesses may have specific needs and conditions that require additional considerations.

- *Ensure board commitment.* Secure support from the board to adopt BIM implementation. Clarify the business's policy and strategy towards BIM. Maintain a clear long-term vision of the benefits of digitalisation, including driving efficiencies, increasing productivity and mitigating disputes with stakeholders.

- *Standardise processes.* Standardise not only BIM documentation but also the working methods of each department affected by BIM implementation. Consistent processes across the organisation promote efficiency and collaboration.

- *Collaborate with the right supply chain.* Identify and collaborate with supply chain partners who share a similar ethos and are willing to embrace digital construction projects. Build strong partnerships and work closely with these companies.

- *Cultivate a learning culture.* Promote awareness of the benefits of BIM and encourage teams to embrace them. Provide training and knowledge to handle challenges during implementation. Foster open and honest conversations about lessons learned, to improve and avoid costly mistakes. Obtain continuous feedback from the teams.

- *Communication.* Provide the right amount of communication and target the recipients. Offer continuous updates but be cautious, as too much information can create fatigue and might produce the opposite of your intended results.

- *Understand the issues.* Identify the challenges that teams face and ensure the correct deployment and support for the successful implementation of new tools.

- *Develop and retain talent.* Invest in the development of your employees and equip them with the necessary knowledge to meet business needs. Address resistance from those who are hesitant to adapt and grow as professionals.

- *Embrace the UK BIM Framework.* Align your processes and documentation with the standards, guidelines and best practices outlined in the UK BIM Framework. Clearly identify project requirements from the tender stage to address vague information and requirements effectively.

Starting the implementation of BIM requires commitment from top leadership, standardisation of processes, collaborative partnerships, a learning-oriented culture, talent development and adherence to recognised frameworks and standards. By following these tips, businesses will be better equipped to embark on a BIM journey.

When is BIM applicable?

The UK BIM Framework is applicable to any project where the appointing party has shown an interest in using it and has supplied the relevant documentation during the tendering process. In this situation, the prospective lead appointed party must clearly specify how they intend to meet BIM expectations as part of their tender response.

It's important to emphasise that the implementation of BIM should be tailored to meet the specific needs of the client and the individual project. This flexibility ensures an efficient and customised approach, allowing for the most effective use of BIM.

As mentioned in Section 2.4, my recommendation is to proactively apply the UK BIM Framework to all projects, unless the client has specifically asked otherwise. The sole exception would be if the client has already progressed to RIBA Stage 3 design using traditional 2D methods, without following BIM principles. For clarification, there is no project value threshold or specific project type that dictates when BIM should be implemented. Its principles are universally applicable to any construction project.

While many countries, including the UK, require the use of BIM for government-funded projects, it is not uncommon for some managers of publicly funded projects to overlook this stipulation. Given this context, I suggest that companies should proactively prepare their teams to be BIM-ready. This will allow them to benefit from its advantages as quickly as possible, irrespective of any government mandates. Companies should not rely solely on external mandates to drive business transformation but should instead take the initiative to unlock the wide array of benefits that digital construction can bring to their operations.

Does the implementation of BIM compromise control, owing to excessive transparency?

This question has come up repeatedly in discussions I've had with construction professionals from various organisations. There is a concern, particularly from a commercial standpoint, that implementing BIM processes leads to a loss of control because of increased transparency – especially with regard to the bill of quantities. Some worry that this level of openness might render the company less competitive or diminish profits.

The enhanced transparency facilitated by BIM has the potential to benefit all parties involved in a project. From a commercial point of view, transparency doesn't entail disclosing proprietary information; rather, it means being open about the project's progress, costs and challenges. Embracing transparency can help in identifying areas for improvement, streamlining processes and building trust with clients and partners. Furthermore, transparency highlights a company's expertise and reliability. Openness and a high level of integrity build confidence and bolster the business's reputation, often leading to repeat business.

So long as all team members execute their tasks with due diligence, transparency should not obstruct competitiveness or give cause for concern about the BIM process. On the contrary, it encourages accountability, fosters innovation and provides motivation.

Is the UK BIM Framework applicable outside the UK?

Yes, the UK BIM Framework is applicable beyond the UK, as it offers guidance and best practices for information management in line with the ISO 19650 series (BSI, 2019, 2020a, 2020b, 2021, 2022). This makes it suitable for implementation in any country. While individual countries may have their own national annexes to adapt the ISO 19650 series to local conditions, the core principles and overarching approach of the UK BIM Framework remain globally relevant and applicable. Its alignment with the ISO 19650 series ensures a standardised approach to managing information in the construction sector, in this manner promoting consistency and efficiency across international boundaries.

How can we improve team engagement and collaboration?

Team engagement and collaboration are vital for driving the change needed in any organisation. It's crucial to break down the barriers that often exist between departments and senior leaders. This issue isn't confined to the digital department; it's prevalent across various departments in most organisations. Fostering interdepartmental collaboration can unlock new potential, leading to innovative solutions and improvements. However, this requires more than just verbal commitments; it calls for decisive action.

One effective tactic is team shadowing within the organisation. This practice not only offers valuable insights into the daily challenges faced by various departments but also demonstrates a genuine interest in the work of different teams. This can build strong relationships and enhance your visibility as an actively engaged member of the organisation, giving you a better understanding of how you can support your colleagues

Senior leadership has a fundamental role to play in supporting the change needed. Merely listening to the concerns and needs of teams won't yield results unless it is followed by concrete actions. Employees are savvy enough to distinguish between empty rhetoric and actual commitment. As such, leaders should explicitly state how they intend to create an engagement and collaboration environment, rather than using overused jargon.

For instance, instead of ambiguously stating, 'We will engage and collaborate,' senior leaders should list the specific steps they plan to take. These could include rigorously adhering to company policies, holding regular open forums to discuss the implementation of digital strategies within the business or holding teams accountable for delivering on their new responsibilities. Actions do speak louder than words, and a genuine commitment to engagement and collaboration will manifest itself in the culture, productivity and, ultimately, success of the business.

9.3.   Chapter 3: Supply chain

If some members of our supply chain haven't yet adopted BIM, does that rule them out from bidding on our projects?

As discussed in Chapter 3, the selection of supply chain partners should not be based solely on cost; their ability to meet project requirements is more critical. It's increasingly important to collaborate only with companies that support your BIM values and methodologies. For certain trades, such as mechanical, electrical and structural, meeting this criterion is non-negotiable. Partnering with a firm that lacks the necessary expertise can jeopardise both the project and your business.

It's important to recognise that not every member of our supply chain has the same responsibilities and requirements when it comes to BIM. Although members must be capable of sharing information and collaborating within a common data environment, they should also adhere to agreed workflows and naming protocols. Crucially, it's necessary to determine which partners are required to produce designs in a 3D environment to support design coordination and to meet the client's expectations.

If a company fails to meet the BIM criteria, it might not be suitable for BIM-specific projects. However, it could still be considered for projects with different requirements.

You also have a responsibility to guide your supply chain partners through digital transformation, to achieve shared business objectives. This support could include clear communication of expectations and even providing workshops to enhance skills. While aiding companies in their BIM journey is important, there comes a point when clear commitments must be established for those wishing to remain part of the supply chain. Ensuring that you partner with capable businesses is crucial to avoid potential setbacks in your projects.

Are the BIM requirements specified in the supply chain appointments?

It is essential for BIM requirements to be specified in supply chain appointments, as outlined in Section 6.6. The success of your projects largely depends on the performance of your supply chain partners; therefore, it's crucial that you perform thorough due diligence during the tender stage and finalise appointments with appropriate documentation to hold your supply chain partners accountable for delivering the project based on the specified BIM requirements.

During the tender stage, it's important to work closely with your supply chain. Ensure you provide all the necessary documentation promptly and address any concerns. Being accessible and approachable for queries is key.

Make sure that all mandatory documentation forms part of the appointment. Supply chain teams must comply with the set requirements; maintaining a close relationship is essential for holding them accountable for high-quality delivery and adherence to the documentation outlined in the appointment.

Should we agree to pay additional fees to the supply chain in order to meet BIM requirements?

It's important to note that not every supply chain increases its fees to accommodate BIM requirements; in fact, this practice is starting to become less common. However, some companies do choose to increase their charges; in this book, I've explored the various reasons why they might do so.

I believe that increasing fees to meet BIM requirements should generally be unacceptable, except in specific instances where there is an extraordinary amount of asset data to deliver. Companies can benefit from the efficiencies that BIM offers, and I think the industry should scrutinise any additional fees imposed by the supply chain for BIM implementation. In fact, some companies that I have interviewed in the past stated that their fees for BIM projects are actually lower than for non-BIM projects.

Supply chain teams should not view BIM as an excuse to raise fees. Instead, all stakeholders should collaborate to realise the benefits that BIM can bring to their projects and, importantly, to their businesses. Meeting BIM requirements should not be seen as an extra service or a favour to clients; it should be considered the standard way of working to enhance productivity and maintain a competitive edge.

How is the adoption of digital tools among task teams progressing?

Over the years, I've seen major improvements in how digital tools are used to help plan and execute projects. Nowadays, most consultants mainly use 3D environments for their design tasks. They've also embraced the common data environment for sharing information, as well as new tools for creating visualisations and coordinating efforts. However, these skills are usually found among just a few people within a company.

Despite this progress, challenges still exist, particularly among smaller subcontractors. Some have yet to realise the benefits of employing digital tools and managing information effectively. This delay could be due to a lack of training resources or the initial financial investment required to make the transition to a more digital workspace. Additionally, while some employees may have the necessary expertise, the overall skill level within the company might not be standardised. This issue is similar to what we see among consultants, as mentioned previously. Such inconsistency leads to a reliance on specific individuals to fulfil project requirements, creating varied experiences, depending on who is involved.

In summary, the industry has made considerable progress, with design consultants and larger subcontractors generally keen to adopt digital tools. However, there is still a need for smaller companies to acquire these skills to qualify for projects with BIM requirements and support the transformation of the industry.

9.4. Chapter 4: Artificial intelligence and data analytics

Can AI potentially replace human roles in the construction industry?

Artificial intelligence has the potential to automate certain tasks in the construction industry. However, in my view, AI is unlikely to fully replace human beings. Instead, it will enhance productivity for those who adopt AI in their workflow. This means that companies could accomplish more work with the same number of employees. Proper use of AI by teams can increase efficiency, reduce errors and improve most aspects of a project's lifecycle.

Is the integration of AI in the construction industry a tangible reality or is AI merely a buzzword?

In the construction industry, AI is more than just a buzzword; the integration of AI is a tangible reality that is already enhancing design and construction processes, as covered in Chapter 4.

It's true that we're currently in a phase where everyone is talking about AI, often selling it as a universal remedy for all challenges within our industry. Referring to Figure 9.1, which shows the Gartner hype cycle, I believe we are currently in the 'peak of inflated expectations' phase (Gartner, 2023). While I expect that the excitement around AI will decrease in the near future, it's evident that AI is already making a positive impact on the industry.

Figure 9.1 The Gartner hype cycle

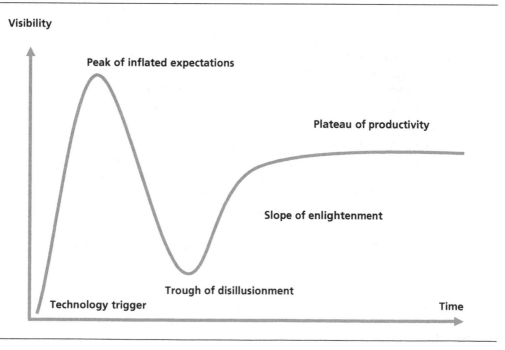

Reflecting on the past, the implementation of BIM in the industry has followed this cycle quite closely.

What considerations would you suggest for implementing AI in this sector?

The following points are intended to guide you as you begin to develop your AI strategy.

- *Strategy and needs assessment.* Begin by delineating your strategy and identifying specific areas within the organisation that could benefit from AI integration. It's imperative to start by pinpointing actual business problems.

- *Engagement with various teams.* Engage with different teams to discover these opportunities, initially adopting a 'bottom-up' approach. This will encourage a more organic understanding of the issues at hand, as opposed to having senior leaders dictate the course of action. However, ensure that your decisions are data-driven, not based solely on opinions.

- *Process optimisation.* Before diving into automation or the implementation of a new system, it is essential to first standardise and simplify your existing processes. Automating a flawed process will only perpetuate inefficiencies. Conduct a thorough review of current processes to analyse the potential benefits of automation or the necessity to remove a process. Consider the different opportunities available, as even minor changes can yield significant long-term benefits.

- *Pilot testing.* Initiate a small-scale pilot project to assess the viability of chosen AI solutions. Focus on low-risk, high-reward tasks for the initial roll-out and ensure you allocate the appropriate resources to support the deployment and assess benefits correctly.

- *Training and development.* Provide your teams with the skills required to engage with the selected AI solutions.

- *Stakeholder engagement.* Involve key stakeholders, such as directors, early in the process. Secure their buy-in by including them in the decision making process. When senior leaders take ownership of the project, you're more likely to receive their full support during implementation.

- *Monitoring and feedback.* Regularly monitor the performance of the AI solution and solicit feedback from end users for ongoing improvements. This is a continuous cycle that involves measuring the success of the implementation, implementing adaptations based on feedback and providing the necessary training. The automation strategy should not only simplify and standardise processes but also allow room for improvements and adjustments.

- *Scale up.* Once the pilot phase is successfully completed and any issues are resolved, you can extend the AI implementation to other areas or projects within the organisation.

- *Review and adapt.* Given the evolving nature of AI, conduct regular assessments to ensure that the solution remains effective and relevant. This process, like training, is a cyclic one that continually assesses benefits and areas for improvement. Note that this strategy should be reviewed every 6 months, owing to rapid changes in this sector. You don't need to overhaul the entire strategy, but minor tweaks will be necessary to keep it current.

9.5. Chapter 5: Understanding BIM and its foundations

Is the BIM Level 2 concept still relevant?

Owing to the BIM Mandate (Cabinet Office, 2011) and the publication of PAS 1192-2:2013 (BSI, 2013), which referenced BIM Level 2, the concept of BIM Level 2 gained widespread popularity within the construction industry. However, this term has often been misused, much like the broader BIM concept. This concept has now been superseded by the UK BIM Framework, which outlines the approach for implementing BIM in the UK, based on the ISO 19650 series (BSI, 2019, 2020a, 2020b, 2021, 2022).

When using the UK BIM Framework, it's crucial not just to reference the framework but also to specify the project requirements clearly. The framework offers a set of guidelines and standards aimed at establishing a consistent and successful approach, thus eliminating the need to start from scratch (UK BIM Framework, 2023). Therefore, the project requirements must be well-defined and not just a reference to the framework, as has occurred in the past – for instance, with BIM Level 2.

It's important to understand that a framework serves as a guide and needs to be adapted to suit the specific requirements of each project. A one-size-fits-all approach doesn't work; flexibility is essential to fit the framework to the unique characteristics and complexities of the project at hand. A thoughtful and tailored approach ensures the successful application of BIM principles and practices, optimising project outcomes.

Are the BIM dimensions still applicable?

Much like the BIM Level 2 concept, the idea of dimensions gained traction around the time the BIM Mandate was implemented. The 3D, 4D and 5D concepts became particularly well-known. However, a large number of other dimensions emerged without a clear consensus behind them. This lack of agreement led to different interpretations, causing unnecessary confusion in the industry.

It's important to understand that these dimensions essentially serve as use cases of how information can be applied throughout a project's lifecycle. That said, it's not necessary to coin a new term for every aspect or try to fit everything into a specific dimension.

Instead of getting caught up in the jargon of dimensions, it's more pragmatic to focus on getting the fundamentals right. Clearly outline your project objectives and desired outcomes in straightforward language within the project's documentation and scope. This approach will help eliminate unnecessary complexity and uncertainty regarding expectations.

While discussions around dimensions continue, my advice is to shift the focus towards directly articulating project requirements and goals. This will promote better communication and understanding among stakeholders, leading to more efficient workflows and successful project outcomes. Ultimately, we should prioritise practicality over jargon and aim for effective collaboration and clarity in our projects.

What is the cost of implementing BIM?

Generally, introducing BIM from the outset does not result in extra costs. It should be considered the industry-standard way of working. Furthermore, the cost savings and benefits gained through BIM make it an obvious choice when the right teams are involved. However, if a request to implement BIM is made midproject, particularly after the spatial coordination stage of the design has already been completed, additional costs are likely to arise. This is because the design team would need to recreate previously developed information to suit the BIM requirements.

While some might argue that point cloud surveys, which are used to verify site conditions or designs, constitute an added expense, I regard them as invaluable. They present an opportunity to mitigate project risks and ensure smoother project delivery. Identifying potential issues early on can help avoid costly rework and delays later in the project.

However, if teams haven't yet started transitioning to BIM or are still in the early stages, additional costs could arise. This is particularly likely if teams' current practices don't align with BIM principles, which could result in duplicated efforts. The situation becomes even more relevant if teams rely on third parties for BIM-related tasks. Transitioning to BIM does involve costs, such as software licences, possible hardware upgrades and training in both the process and the software. Nonetheless, these costs should be seen as investments rather than expenses. The ultimate goal is to deliver projects more efficiently, meet client expectations and secure more work while ensuring successful outcomes. The exact costs will vary depending on the size of the team and its specific needs.

Why minimise the use of folders in the common data environment?

For an effective common data environment (CDE), the focus should be on naming protocols, workflows and metadata, rather than on conventional folder structures. By consistently applying an agreed naming scheme and using relevant metadata, the process of finding and managing information in the CDE becomes significantly easier.

This methodology eliminates the need for subfolders within the CDE, ensuring quick and accurate access to all project information for all team members. This is because relying on folders organised by drawing serial numbers or packages can lead to duplicated or misplaced information. Additionally, each organisation may have a different structure, making it difficult for task teams to familiarise themselves with new systems, especially when working on different projects simultaneously.

The objective of minimising folder use and organising information using metadata and a naming protocol is to establish a more streamlined and efficient manner for managing project information.

If folders must be used, they should be managed very carefully to avoid becoming a tangled mess of nested folders, a common issue in a Windows environment. Even introducing just one level of folders poses a risk of misplacing information. In some cases, teams may struggle to find the information they need. Therefore, my recommendation is to minimise the use of folders as much as possible.

Which takes precedence: the model or the drawings?

It's important to clarify the hierarchy of the project documentation to all stakeholders involved. This will eliminate any ambiguity that could lead to exploitation in the event of disputes.

During the design and construction phases, 3D models serve as tools for design coordination. However, the 2D drawings, which are extracted from these coordinated 3D models and issued with an 'A' code, serve as the contractual documents once they have been authorised and accepted at the end of the workflow.

It's important to note that although the drawings are generated from the federated 3D model, they may contain additional annotations and details not visible in the 3D environment. As such, if a dispute arises, the authorised and accepted drawings should act as the point of reference, taking precedence over the 3D model.

To avoid confusion, it should be explicitly stated that any published information carrying an 'A' code is contractual. Conversely, any information labelled with 'S' codes (S1–S5) denotes a shared status, reflecting the intended use of an information container. This information should not be used on site, as it is non-contractual.

To minimise misunderstandings, conflicts and opportunistic behaviour among stakeholders, the project's BIM documentation should clearly outline the agreed procedures. Open communication and collaboration among team members will further mitigate the risk of discrepancies between the 3D model and the drawings, ensuring a smoother project execution.

What information is shared within the CDE?

Throughout my professional journey, I have utilised the common data environment (CDE) to collect, manage and share project information in compliance with PAS 1192 (BSI, 2013), BS 1192 (BSI, 2016) and, more recently, ISO 19650-2 (BSI, 2021) guidelines. This includes drawings, reports, specifications, point cloud surveys, models and other project-related documents.

However, over time, the adoption of various tools, such as model viewers, task management software, site imagery capture and other solutions, has led to a complex landscape within organisations. This complexity makes it challenging to implement digital solutions, as users must navigate different platforms to find information and take action. Additionally, the administrative tasks required to maintain secure access across these diverse platforms introduce another layer of complexity.

While the CDE serves as a single platform for information exchange, meeting its core objective, the introduction of diverse tools with additional functionalities complicates matters. The fragmentation and duplication of information across different platforms increase the risk of information loss, misuse and challenges in adoption.

To address these issues, businesses should consider streamlining their digital tool architecture for enhanced efficiency and cohesion. The current state of fragmentation calls for a strategic shift.

Fortunately, robust tools are now available in the market that can assist businesses in unifying and optimising project management practices. These tools help mitigate the risk of information loss and enhance operational efficiency. While there is no single tool that covers all business needs, it's essential to establish connectivity between these tools to mitigate the aforementioned risks. This will significantly improve the efficacy of project delivery while ensuring that critical information remains accessible and secure.

Is the content discussed in this book relevant to civil projects?

This book is highly relevant to civil projects. The Infrastructure and Projects Authority (IPA) endorses the importance of accurate information management and use of the UK BIM Frameworks. While my writing is mainly centred on the building sector, owing to my professional experience, the principles and best practices covered apply universally to a range of project types.

In civil projects, design teams might use different technologies to generate information. These technologies may be more suited to the specific needs of civil engineering than those of the building sector. However, the rest of the process remains the same. This includes the challenges and benefits that come from adopting a robust approach to digital construction and BIM.

The shift towards a learning-oriented culture within your organisation, along with the adoption of simplified processes and technology, can lead to accurate information management and collaboration. These changes bring about such benefits as increased productivity and a significant decrease in the cost of error.

9.6.   Chapter 6: Strategy, roles and procurement

Who produces the BIM Execution Plan?

The BIM Execution Plan (BEP) is created by the lead appointed party. However, before the appointment, this party had to provide the pre-appointment BEP as part of the tender response.

The identity of the lead appointed party can change, depending on the current stage of the project and the chosen procurement approach. As covered in the glossary, a lead appointed party is any company that holds a direct appointment from the appointing party. Therefore, some projects may have a number of lead appointed parties. In such cases, these parties must collaborate to ensure that the BEP and MIDP are coordinated, particularly if more than one BEP exists.

I have previously encountered two primary scenarios.

- In projects where a traditional procurement method is used, it's often the architect, acting as the lead designer, who assumes responsibility for creating the BEP. The other design consultants then work in accordance with the BEP produced by the architect. When the main contractor becomes involved in the project, they have two options. One option is for the main contractor to create a new BEP based on the architect's latest version and update it regularly for the duration of the project. A second option is to continue using the architect's existing BEP, provided that the architect takes responsibility for keeping the plan updated throughout the work's execution. In both scenarios, the main contractor is required to include the EIR and all other relevant documentation, as discussed in Section 6.6, as part of the appointment documentation delivered to the new task team responsible for delivering the CDP packages. As always, clear communication with the teams is required, as other approaches may be applicable.

- In a design-and-build project scenario, the main contractor takes on the responsibility of creating the BEP from scratch, collaborating with the known task teams to create the document.

What are the task information delivery plan and the master information delivery plan?

The master information delivery plan (MIDP) has replaced the traditional information release schedule (IRS) to offer a more comprehensive understanding of the expected information from each discipline at every stage. While similar in nature, the MIDP takes into account predecessor information. It identifies the necessary data for the accurate development of drawings and other project documentation, all of which are linked to the project programme.

The MIDP consolidates the task information delivery plans (TIDPs) from all the different task teams. The TIDP replaces the previously used design delivery programme. The TIDP is a key document that lists all project information, including not just drawings but also various details to be issued for each package. This allows the lead appointed party to review and determine whether the proposed drawings are likely to meet project needs and expectations. It provides an opportunity to discuss any additional information or drawings required before the appointment. Moreover, the TIDP helps identify errors in naming protocols, saving time and effort that would otherwise be spent in rejecting and renaming drawings.

The TIDP must be completed as part of the tender stage and is then reviewed and accepted by the lead appointed party. Once accepted, the TIDP becomes part of the appointment. Any changes to it are initiated exclusively through the change control procedure.

While it's true that the TIDP is likely to be reviewed during the project, the task team should, given their professional expertise, be capable of providing a well-approximated TIDP based on the amount of information they expect to produce for the project at tender stage.

Other methods exist for delivering a TIDP, but this approach has proven effective in the past. It allows teams to understand the volume of information to be delivered on specific dates and to adjust those dates accordingly. This ensures the team has enough time for review and feedback, prevents the simultaneous submission of all project information and benefits every party involved.

Who is responsible for uploading information to the common data environment?

From the perspective of the main contractor, when information is received during the tender stage, the information management team will establish and configure the common data environment (CDE) and will upload all relevant information to the CDE to facilitate information management at this stage, particularly if the appointing party has not already done so. This approach enables internal teams to access this information and disseminate it to prospective task teams bidding for the project.

However, after a task team has been appointed, the responsibility for uploading data to the CDE shifts to that specific task team. It is crucial that teams each upload their individual information to the CDE, using the correct naming conventions and metadata, in accordance with the task information delivery plan (TIDP). This ensures that all project stakeholders can access the most up-to-date and accurate information and maintain accountability for its content. Information should not be sent through email with the assumption that the information management team will upload it to the CDE on behalf of the task team.

Once the information has been uploaded to the CDE by the respective task teams, the information management team will conduct the necessary quality assurance checks as part of the approval workflow before it is made available to the rest of the project team.

Who is responsible for federating the model?

The process of federating the model may vary, depending on the tools used. Traditionally, the lead designer has been responsible for combining the design models to facilitate design coordination. However, new technologies can now automate this process, provided that each task team's model shares the same coordinates.

The responsibility for uploading models to the common data environment (CDE) or to the chosen platform for model federation – known as the model viewer in some platforms – rests with individual task teams. Ideally, these platforms should be the same or interconnected.

Don't confuse the federation of models, which enables the delivery team to understand how the different design disciplines are coordinated, with clash detection that uses the federated model and specific technology to identify problems in the design. Depending on the tools selected for the project, these might require lead designers to create their own federated models within the specialised platform. As covered in Section 7.2, it's essential to understand that the lead designer bears the responsibility for overseeing design coordination. This can be done using either the federated model or 2D drawings. However, it is abundantly clear that identifying design mistakes and risks is easier in a 3D environment when using the federated model. Therefore, I encourage all teams to adopt this approach when reviewing design coordination.

It's important to remember that, during the ongoing model revisions issued throughout the project, task teams must each ensure that their design model has the correct coordinates to support design federation. Furthermore, they must complete a design coordination review with other disciplines before sharing design models with the rest of the team.

Clearly defining the responsibilities of each task team and the approach to design coordination during the tendering phase is critical to avoid misunderstandings or assumptions. This clarity ensures that the project runs smoothly, reduces reluctance over coordination throughout the project and minimises errors and issues.

What is the difference between the level of detail, the level of information and the level of information need?

The level of information need, as introduced in the ISO 19650 series (BSI, 2019, 2020a, 2020b, 2021, 2022) and further elaborated in BS EN 17412-1:2020 (BSI, 2020c), serves as a framework that defines the scope and depth of information to be exchanged. This includes various types of data, such as geometrical or alphanumerical data or documentation.

This topic has always been very confusing, owing to the amount of terminology and acronyms used. The terms 'level of detail' (LOD), focusing on graphical content, and 'level of information' (LOI), pertaining to non-graphical content (Swaddle, 2022), have been widely used. However, inconsistencies in definitions and references led to confusion. For example, American standards used different LOD levels (100, 200, 300, etc.) from those used in the UK (LOD 2, LOD 3, LOD 4). Additionally, LOI was sometimes mistakenly equated with the requirement to deliver COBie data, which was not its original purpose. Moreover, such terms as level of development and level of definition were introduced, which blend aspects of both LOD and LOI. This amalgamation added an extra layer of complexity and confusion to the topic.

As a result, although still being referenced, the industry is gradually moving away from the terms LOD and LOI, which are not referenced in the ISO 19650 series. The level of information need is a broader framework that extends beyond the traditional focus of the LOD and LOI.

The shift to the level of information need has generated a variety of reactions. Some see it as a step in the right direction, while others find it confusing. What's most important is that the content of the deliverables should meet expectations precisely – neither exceeding them nor falling short. The framework offers guidance to communicate these expectations effectively to various stakeholders at different stages of the project, ensuring accountability. Adopting a consistent terminology makes it easier for everyone involved to understand and meet a project's requirements.

The responsibility for determining the level of information need for each information deliverable lies with the appointing party, who sets the method for defining it as part of the project's information standard, but, if lacking in the necessary expertise, can consult the lead appointed party or an independent third party for guidance.

This framework facilitates the communication of essential information based on its intended use and helps in specifying the requirements for each information container, thereby avoiding both excessive and insufficient information.

9.7. Chapter 7: Advanced topics and best practices

Compared with new construction, do the challenges of using BIM differ for refurbishment projects?

Refurbishment projects naturally have unique challenges and risks compared with new construction projects. Introducing BIM to existing buildings brings several additional considerations that might not be relevant for new constructions. For example, the appointing party often lacks comprehensive data about a building's current state, which are essential for advancing the design process and making decisions. Limited access to conduct the necessary surveys is a common challenge for teams working on existing buildings. Furthermore, it's important to consider existing services, as they can significantly affect the design of new additions – especially in projects where some services must be retained.

Nonetheless, BIM can streamline the management of refurbishment projects and reduce associated risks. By using point cloud surveys and managing information effectively within a 3D environment, the design team can gain a clearer understanding of the building's current condition. This enables them to make more informed decisions, enhance collaboration, reduce rework on site and provide the appointing party with more precise data for asset management.

However, the challenges highlighted are not exclusive to the implementation of BIM. These challenges arise in any project dealing with an existing facility, even when traditional methods are used. Therefore, in response to the question, the additional challenges are not attributable to BIM but rather to the complexities of working with existing buildings.

In situations where surveys cannot be carried out for an existing building refurbishment project, what approach should be taken?

If surveys cannot be conducted for a refurbishment project of an existing building, owing to limited access before the design phase starts, it's essential to communicate the associated risks to the client. This limitation often arises when a lease term has not yet finished, preventing any form of work on site. Relying on existing documentation to develop the design introduces a significant risk. Often, the existing documentation doesn't accurately represent the current site conditions, leading to numerous assumptions that could affect the work later on.

Therefore, it's crucial to account for this risk during the tender stage and allocate sufficient time in the project programme to conduct the necessary surveys as soon as access is granted. This will allow you to verify and update the design to ensure its accuracy before commencing any new work. In cases like these, conducting a post-strip-out survey is usually the best approach to verify the available design.

This requirement is applicable whether the project uses BIM or not. The absence of accurate surveys means that you'll be relying on assumptions, which can lead to problems during the construction phase. If access to the site cannot be allowed before the design phase begins, it must be secured before the installation of new elements on site, to verify the design and rectify any errors arising from initial assumptions.

Some projects might benefit if surveys are conducted in different zones, to expedite the verification process and allow work to start on site sooner. However, the main takeaway is to verify the design as promptly as possible. Understanding the impact of inaccuracies and proactively updating the design can help mitigate potential issues before they escalate into major problems during construction.

How does the level of detail change during the various stages of a project?

The expectation that the 'level of detail' (LOD) of elements within the model will change during different project stages is not something I've personally encountered. This can be confusing for some and is often a point of dispute during the tender stage, as different task teams might have varied interpretations of LOD requirements, leading to some anxiety.

For instance, I've never seen an architect use a door object at Stage 2 and then switch it out for a more detailed one at Stages 3 or 4. While the architect might adjust the dimensions of the door, that's usually the limit of any change. If the architect is happy with an object and it has the correct level of detail for the purpose of producing Stage 4 information, there's no need to oversimplify the object for previous stages.

As a project progresses, the design model might require additional objects not needed in earlier phases. These additional objects enhance the model's quality, supporting various stages of design for coordination and information production. This should not be confused with the level of detail (LOD) within each object. It's essential to distinguish between the specific geometry of an object at a particular stage and the overall number of objects required to comprehensively describe the design. The latter is likely to increase as the project advances through different stages.

When considering the LOD, the primary factor should always be the intended purpose. For instance, steel subcontractors creating a design model will naturally include all required elements with high precision, as this is essential for generating the proper documentation and manufacture.

A problem around this topic is that there are instances where users download objects with an unnecessarily high level of geometric complexity. This can cause issues later in the project. Such objects may be suitable for rendering and visualisation but are not practical for design. Therefore, be cautious and make sure that your team avoids downloading uncontrolled copies. If downloading is unavoidable, ensure that the object meets all discipline requirements and doesn't include irrelevant details. In these scenarios, teams may need to have different sets of libraries: one set of objects for the purpose of visualisation and another set of objects for design purposes.

In summary, my view has always been that design models should contain all the necessary details and information required at each stage, to allow production at the expected quality.

Are BIM tools more efficient than 2D CAD tools?

The *CAD and BIM Productivity Study*, published by Autodesk but independently conducted by David Cohn (2015), provides valuable insight into the productivity advantages of BIM tools over traditional 2D CAD tools. In the study, the efficiency of AutoCAD 2015 and Revit 2015 were specifically examined, comparing their effectiveness in creating construction documents for a hospital extension. Although the setup times for both tools were similar, Revit clearly outperformed AutoCAD in terms of productivity. Revit quickly generated various design views, and design changes were automatically updated across all related documents. By contrast, AutoCAD required tedious manual updates. To illustrate, AutoCAD took 48% longer than Revit to complete identical tasks, equating to an impressive 91% productivity increase with Revit.

These findings emphasise the significant potential of BIM tools to improve the efficiency of design teams. The considerable advantages of using Revit should encourage architectural firms to reconsider their current design methods and technologies.

Although the report was conducted in 2014, it remains a valuable reference for anyone questioning the effectiveness of BIM. I believe that the productivity gap between BIM and traditional 2D CAD has increased in recent years, thanks to advances in automation within the BIM environment.

In summary, there's compelling evidence to support the adoption of BIM tools. I understand that some people might be sceptical about the report's findings; however, I'm confident that once you've learned to work with BIM tools, you won't want to return to a 2D environment. At least, that has been my experience.

Please note that the tools mentioned in the report are just examples of BIM tools. The market offers a wide variety of options, and it's your responsibility to conduct appropriate due diligence to select the most suitable tool for both your business and your personal development.

When will digital twins become the norm?

Achieving clarity and structure in our information management processes is vital for navigating the complex yet exciting world of digital twins. Before delving into these advanced technologies, it's crucial to first master the basics of BIM, ensuring that the information management framework has been implemented to a point that is reliable, accurate and efficient.

The future of digital twins holds enormous promise, with the potential to revolutionise not only the construction industry but various other sectors as well. However, as I've discussed in this book, the efficient adoption of BIM remains a challenge and has yet to gain widespread acceptance in our industry. Therefore, although the idea of moving to digital twins is appealing, it might still be some distance away for the majority.

While digital twins, the Internet of Things (IoT) and artificial intelligence (AI) often capture the spotlight, my personal focus is more on foundational aspects. We still need to develop the appropriate skills for managing information within project teams. Moreover, we require clients who are aware of the benefits of asset data and can clearly articulate their requirements for their assets.

Discussing digital twins and other cutting-edge technologies is certainly worthwhile. But it's important not to let these discussions become a costly distraction from the basics. The priority should still be to enhance our information management practices, ensuring that the industry becomes proficient in these fundamental aspects before moving on to more advanced technologies.

By adopting a phased approach, we can improve our BIM skills, setting the stage for the successful integration of digital twins and other advanced technologies in the near future.

What are the implications of printing drawings?

The printing of drawings carries various implications, particularly concerning the maintenance of up-to-date project information, especially during the construction stage. There's a risk that the information within the common data environment (CDE) might be revised after drawings have been printed, making the printed version outdated. This can lead to on-site teams unknowingly working from incorrect versions, thereby causing expensive rework and delays in the project programme. As a result, I advocate that teams utilise the CDE to access project information and ensure that they always work with real-time updates to project data.

I understand that, in certain circumstances, such as adverse weather conditions, or for specific tasks, teams might find printed drawings more convenient. In these instances, it's crucial to ensure that the printed drawings are the most current versions before using them and never to assume that the printed drawings are still valid. Some CDE platforms now offer a quick-response (QR) code feature, allowing teams to quickly verify the drawing's currency. This can serve as a practical workaround to ensure that a drawing is still relevant, if there is a specific need for printing.

Similarly, during the design phase, some teams still prefer to print out drawings to manually add comments and mark-ups. In the best-case scenario, these are then scanned and uploaded to the CDE. While this method might seem convenient for some, it poses challenges in tracking the status and progress of the comments within the drawings. It is difficult to trace manually added comments and confirm that they've been addressed before authorising the information container. Therefore, it's advisable to transition to a digital process for reviewing and commenting on drawings or other information. Modern CDE solutions allow the project team to create comments and markups directly on the information available in the CDE, offering a more user-friendly approach. This method ensures better traceability of comments and is significantly more time-efficient, making it a superior use of the team's time compared to the process of printing to create comments, scanning, and uploading the information to be distributed.

Although there might be an initial learning curve, the long-term advantages are substantial. Digitising the process enhances traceability, minimises risks associated with outdated information and increases productivity. Over time, the merits of adopting digital methods over traditional ones will become evident to all stakeholders, leading to more efficient and error-free project management.

BIBLIOGRAPHY

BSI (2013) PAS 1192-2:2013: Specification for information management for the capital/delivery phase of construction projects using building information modelling. BSI, London, UK.

BSI (2016) BS 1192:2007+A2:2016: Collaborative production of architectural, engineering and construction information. Code of practice. BSI, London, UK.

BSI (2019) BS EN ISO 19650-1:2018: Organization and digitization of information about buildings and civil engineering works, including building information modelling (BIM). Information management using building information modelling. Part 1: Concepts and principles. BSI, London, UK.

BSI (2020a) BS EN ISO 19650-3:2020: Organization and digitization of information about buildings and civil engineering works, including building information modelling (BIM). Information management using building information modelling. Part 3: Operational phase of the assets. BSI, London, UK.

BSI (2020b) BS EN ISO 19650-5:2020: Organization and digitization of information about buildings and civil engineering works, including building information modelling (BIM). Information management using building information modelling. Part 5: Security-minded approach to information management. BSI, London, UK.

BSI (2020c) BS EN 17412-1:2020: Building information modelling. Level of information need. Concepts and principles. BSI, London, UK.

BSI (2021) BS EN ISO 19650-2:2018 & Revised NA: Organization and digitization of information about buildings and civil engineering works, including building information modelling (BIM). Information management using building information modelling. Part 2: Delivery phase of the assets. BSI, London, UK.

BSI (2022) BS EN ISO 19650-4:2022: Organization and digitization of information about buildings and civil engineering works, including building information modelling (BIM). Information management using building information modelling. Part 4: Information exchange. BSI, London, UK.

Cabinet Office (2011) *Government Construction Strategy*. Cabinet Office, London, UK. https://assets.publishing.service.gov.uk/media/5a78ce8eed915d07d35b2933/Government-Construction-Strategy_0.pdf (accessed 23/11/2023).

Cohn D (2015) *CAD and BIM Productivity Study*. Autodesk, San Francisco, CA, USA.

Gartner (2023) Gartner Hype Cycle. https://www.gartner.co.uk/en/methodologies/gartner-hype-cycle (accessed 02/12/2023).

Swaddle P (2022) Level of Detail (LOD) and Digital Plans of Work. https://www.thenbs.com/knowledge/level-of-detail-lod-and-digital-plans-of-work#:~:text=LOD%20%3D%20'level%20of%20detail','%20(non%2Dgraphical) (accessed 04/12/2023).

UK BIM Framework (2023) The overarching approach to implementing BIM in the UK. https://www.ukbimframework.org (accessed 24/11/2023).

Amador Caballero
ISBN 978-1-83549-446-2
https://doi.org/10.1680/iceedc.9446210
Emerald Publishing Limited: All rights reserved

Index

Printed in the USA
CPSIA information can be obtained
at www.ICGtesting.com
JSHW051349130624
64751JS00021B/433